URANIUM

TKS MURTHY

INDIA • SINGAPORE • MALAYSIA

Notion Press

Old No. 38, New No. 6
McNichols Road, Chetpet
Chennai - 600 031

First Published by Notion Press 2019
Copyright © TKS Murthy 2019
All Rights Reserved.

ISBN 978-1-68466-507-5

This book has been published with all efforts taken to make the material error-free after the consent of the author. However, the author and the publisher do not assume and hereby disclaim any liability to any party for any loss, damage, or disruption caused by errors or omissions, whether such errors or omissions result from negligence, accident, or any other cause.

No part of this book may be used, reproduced in any manner whatsoever without written permission from the author, except in the case of brief quotations embodied in critical articles and reviews.

Contents

1.	From Fantasy to Reality	1
2.	Uranium and Radioactivity	12
3.	Uranium and Nuclear Fission	23
4.	Uranium Prospecting and Mining	34
5.	Nuclear Fuel Production	49
6.	Uranium-235 Enrichment	66
7.	Nuclear Power Reactors	84
8.	India's Nuclear Power Reactor Programme	103
9.	Spent Nuclear Fuel Reprocessing	115
10.	Safety in Nuclear Industry	126
11.	Spent Nuclear Fuel Management and Reactor Decommissioning	153
12.	Nuclear Power and Public Perception	164

Additional Reading *167*

1
From Fantasy to Reality

1.1 H.G. Wells' Prophecy on Uranium Power

Fredrick Soddy who collaborated with Ernest Rutherford in the classic studies on the phenomenon of radioactivity, in his book *The Interpretation of Radium* (John Murray, 1909) was the first to suggest that "the most significant application of atomic science would be in providing mechanical and electrical power." He imagined an atom-powered humanity with the ability to "transform a desert continent, thaw the frozen poles and make the whole world a smiling Garden of Eden." Drawing inspiration from this observation, H.G. Wells (1866–1946), known as "The Father of Science Fiction," in his book *The World Set Free* (Macmillan, 1914), wrote: "The problem, which was already being mooted by scientific men like Ramsay, Rutherford, and Soddy, in the very beginning of the twentieth century, the problem of inducing radioactivity in the heavier elements and so tapping the internal energy of atoms, was solved by a wonderful combination of induction, intuition, and luck by Holsten (an imaginary professor) so soon as the year 1933." In the words of Holstein

> *"This little bottle contains about a pint of uranium oxide, that is to say about fourteen ounces of the element uranium. It is worth a pound, ladies and gentlemen, in the atoms in this bottle there slumbers at least as much energy as we could get by burning a hundred and sixty tonnes of coal. If at a word, in one instant, I could suddenly release that energy here and now it would blow us and everything about us to fragments; if I could turn it into the machinery that lights this city, it could keep Edinburgh brightly lit for a week. But at present no man knows, no man has an inkling of how this little lump of stuff can be made to hasten the release of its store."*
>
> *"Given that knowledge mark what we should be able to do! We*

should not only be able to use this uranium and thorium, not only should we have a source of power so potent that a man might carry in his hand the energy to light a city for a year, fight a fleet of battleships, or run one of our giant liners across the Atlantic; but we should also have a clue to quicken the process of disintegration in all other elements…"

"It would mean a change in human conditions that I can only compare to the discovery of fire, that first discovery that lifted man above the brute."

The power from a chain reaction would wipe out everything for miles. Wells further forecasted that a bomb (calling it for the first time 'the atomic bomb') of enormous destructive capability would first explode in 1956. Leo Szilard is reported to have been inspired by Wells' book and helped him to experimentally establish the nuclear fission chain reaction that led to the assembly of the atom bomb.

These predictions by Wells were very startling, considering at the time of the publication of his book the investigations on the phenomenon of radioactivity of uranium and thorium were only at a rudimentary stage, and uranium with its limited industrial applications was only a laboratory curiosity.

1.2 Discovery of Uranium

The story of uranium began during the early years of the sixteenth century at Joachimsthal in Czechoslovakia. The region attracted a large mining population thanks to the discovery of silver in the area. The silver extracted was used for the minting of roughly two million large silver coins called Joachimsthaler, later known more simply as Thaler. These silver Thalers became an accepted currency worldwide, and the word 'Thaler' over the course of time became 'dollar' in the English-speaking world.

While mining for silver, the miners found a shiny black mineral in the mines, which was of little interest them. They nicknamed it 'pitchblende,' a German word essentially meaning 'bad-luck rock.' More than a century later, this mineral attracted the attention of Martin Heinrich Klaproth. He carried out a complete analysis and in 1789, he isolated a strange kind of 'half metal' from it. When added to glass, this substance created vibrant yellows and greens. He named it 'uranium' after the planet Uranus that had been discovered a little earlier. What Klaproth isolated was actually the oxide of uranium. Pure uranium metal was first prepared in 1841 by the French chemist Eugène Peligot. For about a century and a half after Klaproth's discovery, the main application for the metal oxide was for producing yellow glass exhibiting green fluorescence, and glazes for ceramics and porcelain. Uranium stained glass was used for fabricating an incredible number of objects such as lamps, jewelry, plates, bottles, drinking glasses, vases, buttons, cups etc. Oblivious to the hazards of the radioactivity associated with the element, the industry continued to use it well into the twentieth century.

During the next hundred years, geological deposits of uranium less rich than Joachimstahl deposits were discovered in Cornwall in Britain, Moravia in France and in Austria and Romania.

1.3 Radioactivity of Uranium

The last decade of the 19th century saw path-breaking discoveries of the subatomic nature of matter leading to the unveiling of the phenomenon of radioactivity of uranium by Henry Becquerel in 1896. With this discovery, uranium attracted great interest in the scientific world. The ionisation chamber developed by the Curie couple (Pierre and Marie Curie) for the quantitative measurement of radioactivity helped Marie Curie conclude that the mineral pitchblende contained, in addition to uranium, another element more radioactive than uranium. Starting with a few tonnes of the pitchblende mineral secured from Joachimstahl, Marie Curie isolated a fraction of a gram of this new element radium named radium bromide after two years of exhausting work. Another element polonium was isolated from the pitchblende. It was later established that these elements were products formed through the radioactive transformation of uranium. Demand for the element radium grew with its use in luminous dials and also with claims for its efficacy in curing some diseases, especially cancer. Allured by its curative properties many people in Europe and America began using radium in many undesirable ways such as an additive in chocolates and beer! To meet the demand many countries started extraction of the element from uranium minerals. Apart from Czechoslovakia, other countries that entered the field were France, UK, and Portugal. Through exploiting its carnotite deposits in Colorado, the US secured a monopoly of radium-uranium industry in the first decade of the 20th century.

While these developments were taking place, the world became aware of the dangers of exposure to radiations from radium and other radioactive products. Many workers engaged for several years in the Joachimsthal mines were haunted by a mysterious disease which was called "mountain disease." With lungs badly damaged, they suffered from a persistent cough and were spitting blood, a typical symptom of lung cancer. In the US, there was an increased incidence of cancer among the girls employed in painting the luminous dials with radium-containing paints. The Curies who handled the products unaware of the risks were ill with radiation sickness. Ultimately, Marie Curie died of pernicious anemia due to overexposure to radiation.

With the discovery of rich deposits of uranium minerals at Shinkolobwe in the Belgian Congo, which was of a higher grade than ever found anywhere else in the world, the radium extraction activity shifted from the US to Europe in 1915. The mining rights at Shinkolobwe were vested with the Belgian mining trust, Union Minière du Haut Katanga, which was already mining the rich resources of copper and cobalt in the region. After the First World War, a factory at Olen near Antwerp in Belgium began production of radium from African pitchblende. The amount of radium extracted at various plants, predominantly in the Olen plant, was estimated at one and a half kilograms. With the demand for radium slowing down, the Congo mine was closed in 1937, by which time more than 2000 tonnes of the ore containing 65 percent U_3O_8 were stockpiled in the mining zone. Speculating on its future demand, Edward Sengier, the Director of Union Minière, transferred this stockpile clandestinely from the Congo to the US in mid-1940 and stored it in a warehouse in Staten Island, New York. Germany took away about 1200 tonnes of uranium in the form of uranium compounds stored in the Olen refinery when it invaded Belgium in 1940.

1.4 US Embarks on the Atom Bomb Project

The pioneering studies in radioactivity, initiated by Madame Curie and pursued by Rutherford, Soddy, Chadwick, the Joliot-Curies, Frisch, Meitner, and Fermi led to the discovery of the fission of the uranium nucleus in 1939 by Hahn and Strassmann. Most significant was the fact that in terms of Einstein's law of mass-energy conservation the fission reaction releases huge quantities of energy through annihilation of matter. The rare isotope of uranium U-235 present in natural uranium at a concentration of about 0.72 percent was found to undergo this fission process with thermal neutrons. Nuclear fission as a potential source of unlimited energy both for destructive (atom bomb) and constructive (electrical power production) purposes attracted worldwide attention, setting the scene for uranium to become the major actor.

There were several political developments at the international level during that period. Adolph Hitler rose to power in Germany. His anti-Semitic policies drove several nuclear scientists including Albert Einstein, Otto Frisch, Lise Meitner, Leo Szilard, Edward Teller, and Eugene Wigner out of Germany. Some of these scientists moved to the US and were later joined by Niels Bohr from Denmark and Enrico Fermi from Italy. They worried about the potentially disastrous consequences if Germany, which was under Hitler, secured a lead in the assembly of the atom bomb. They decided that the US should take immediate steps to stall Germany. Leo Szilard took the lead, approached Albert Einstein and persuaded him to address a letter to President Roosevelt on the urgent need for the US to act on the matter. The following are excerpts from the letter dated August 2, 1939, written by Albert Einstein addressed to President Roosevelt.

"Some recent work by E. Fermi and L. Szilard, which has been communicated to me in manuscript, leads me to expect that the element uranium may be turned into a new and important source of energy in the immediate future. Certain aspects of the situation which have arisen seem to call for watchfulness and if necessary, quick action on the part of the administration. I believe, therefore, it is my duty to bring to your attention the following facts and recommendations.

——— It may be possible to set up a nuclear chain reaction in a large mass of uranium, by which vast amounts of power and large quantities of new radium-like elements would be generated.

―――― This new phenomenon would also lead to the construction of bombs, and it is conceivable – though much less certain – that extremely powerful bombs of a new type may thus be constructed. A single bomb of this type…exploded in a port, might very well destroy the whole port together with some of the surrounding territory.

―――― You may think it desirable to have some permanent contact maintained between the administration and the group of physicists working on chain reaction in America.

I understand that Germany has actually stopped the sale of uranium from Czechoslovakia, which she has taken over."

President Roosevelt responded by initiating a programme to explore the military applications of nuclear energy and constituting an Advisory Committee on uranium headed by Lyman J. Briggs, Director of the National Bureau of Standards. The Advisory Committee recommended that the US Government fund limited research on uranium isotope separation as well as on fission chain reaction at the Columbia University. On September 1, 1939, Germany invaded Poland. Britain, France and several countries declared war against Germany.

In June 1940, the President merged the Advisory Committee on uranium with the newly created National Defense Research Committee (NDRC) under Vannevar Bush, an engineer, and able administrator. With security in mind, Bush barred foreign-born scientists from the committee and banned publication of articles on uranium research.

1.5 First Atomic Pile

Enrico Fermi and Walter Zinn who were conducting studies on uranium fission reactions at Columbia University predicted that a self-sustaining nuclear fission pile is possible if a sufficient quantity (critical mass) of uranium is assembled in a suitable configuration with graphite as a moderator. But to be effective the graphite has to be pure, nearly free from neutron-absorbing impurities like boron. The NDRC transferred the work on the nuclear pile to Chicago University, where Fermi's team which included Szilard, conducted as many as 30 sub-critical pile tests and arrived at the correct critical configuration of uranium metal in a matrix of graphite.

Work on the experimental pile started on November 16, 1942. A squash court beneath the west stands of a football field (Stag Field) was chosen for its location. Material procurement, however, caused problems. Though Fermi's team preferred uranium metal as the fuel in the pile, only about six tonnes of metal of required purity could be made available by Frank Spedding of the Iowa State University. Without waiting for the required quantity of the metal, the team decided to use about 40 tonnes of uranium oxide. About 380 tonnes of precisely machined blocks of graphite were used as a moderator. The pile, which was square at the bottom and a flattened sphere at the top, was enclosed in a huge balloon specially made by Good Year Tyres Co. to quarantine the gases released during the fission reaction. The chain reaction was controlled by cadmium rods introduced in the pile for absorbing the neutrons. By December 1942, the pile was ready for testing. With 50 scientists and technical experts, led by Fermi and Szilard and the laboratory's health and protection staff closely monitoring the procedure, the first ever nuclear chain reaction in a pile was demonstrated at 3:53 PM on December 2, 1942. With a multiplication factor of 1.006, the neutron chain reaction was slow enough to be controlled by the manipulation of the cadmium rods, generating a power of just 40 watts, barely enough to light an ordinary incandescent lamp! The success of the experiment heralded a new era in scientific achievement. This reactor was also to become the basic model for all nuclear reactors.

Fig. 1: First Atomic Pile

The success of the pile assembly, which was a crucial part of the atom bomb programme was a tribute to Fermi and his team. By the time, Fermi's pile demonstrated self-sustaining nuclear fission, considerable experimental data was generated in various laboratories in UK and USA, which would help in the bomb project.

Meanwhile, in 1941 G.T. Seaborg and his coworkers isolated the synthetic element plutonium-239 formed in the reaction between slow neutrons and uranium-238. This α-emitting element with a half-life of 24,200 years, which can be separated from uranium is shown to be capable of undergoing fission with energy release. This discovery created additional interest in the atomic pile as a source of the new fissionable plutonium-239 for making the bomb.

1.6 Manhattan Project

Following the attack on Pearl Harbour by Japan on December 7, 1941, and the declaration of war by Germany and Italy on the US, the Project Manhattan Engineer District (MED) was established in

August 1942 in New York for expediting the work on building the atom bombs. The Project, which was to soon be known as the Manhattan Project, was sanctioned the US $2 billion. Two methods were considered for making the weapon, one was based on enriched uranium (uranium-235) and the other, using the plutonium produced in a reactor. It was decided to explore both alternatives. To this end, the Executive Committee recommended building one or two reactors for producing plutonium and pilot plants for U-235 enrichment using electromagnetic, centrifuge and gaseous diffusion technologies. With the news of Germany's nuclear programme trickling in, the tempo of the project was heightened. General Leslie Groves was given command of the project in September 1942. The first reactor, named X-10 for the production of plutonium was set up at Oak Ridge, Tennessee. Hanford, a remote region in Washington State near the Columbia River was chosen for production of plutonium from the irradiated reactor fuel. In October 1942, E.I.du Pont de Nemours agreed to set up a full-scale chemical plant at Hanford to separate plutonium from the irradiated reactor fuel. By November 1942, it was decided to skip the pilot plant steps for isotopic enrichment of uranium and proceed with building full-scale gaseous diffusion and electromagnetic plants at Oak Ridge, Tennessee. Los Alamos, New Mexico, was selected for the bomb assembly project with Robert Oppenheimer as its director. Work on the first Hanford reactor was started in June 1943 and operation began in September 1944. This was followed by construction of more reactors.

The immediate concern of the Manhattan Project was to procure the uranium ore required for the project. The only domestic supply source for the US was the Colorado deposit. But this deposit was already closed due to competition from Union Minière. The Joachimstahl source was not accessible as Czechoslovakia was run over by Germany. At this point, the 1,200-tonne stockpile of uranium ore brought by Edgar Sengier, and stored in the Staten Island warehouse proved a windfall. General Groves sent Colonel K. D. Nichols to buy whatever uranium ore he could from Sengier and the Belgian company. The entire stock of ore at the Staten Island warehouse and an additional 3,000 tonnes of ore stored above ground at the Shinkolobwe mine were purchased by Colonel Nichols for use in the project. In the words of Col. Nichols,

"Our best source, the Shinkolobwe mine, represents a freak occurrence in nature. It contained a tremendously rich lode of uranium pitchblende. Nothing like it has ever been found. The ore already in the United States contained 65% U_3O_8 while the pitchblende Above ground in the Congo amounted to a thousand tonnes of 65% ore, and the waste piles of ore contained two thousand tonnes of 20% U_3O_8. The uniqueness of Sengier's stockpile can be appreciated from the fact that the MED and AEC considered ore containing as low as three-tenths of 1% as a good find. Without Sengier's foresight in stockpiling ore in the United States and above ground in Africa, we simply would not have had the amounts of uranium needed to justify building the large separation plants and the plutonium reactors."

The US Army also sent a squad from its Corps of Engineers to the Congo to restore, at a cost of $13 million, the mines that were closed in 1939. Between 1942 and 1944, about 30,000 tonnes of uranium ore were acquired by the US Army from this source.

By April 1943, work on the bomb design began at Los Alamos. The Manhattan Engineering Division moved to Oak Ridge later in the year. By early 1944, Los Alamos started receiving enriched uranium. Within less than two years of the inauguration of the project, Hanford reactors began producing plutonium. By February 1945, Los Alamos received its first plutonium. Meanwhile, tests were being done on dummy bomb models.

1.7 The Trinity Test: First Ever Nuclear Explosion

As the project started taking shape, the US military remodeled B-29 bombers for the delivery of atom bombs. Harry S. Truman succeeded Roosevelt as the US President, following the latter's death in April 1945.

By June 1945, two bombs were ready. One bomb, based on enriched uranium was named, "Fat Man" and the other based on plutonium, "Little Boy." Some scientists among the Los Alamos group were skeptical about the effective performance of the devices. The Franck Committee recommended conducting a test to assess the destructive potential of the atom bomb before considering its use against Japan. As there was enough weapons-grade uranium for just one bomb while plutonium was available in sufficient quantity for making a few bombs, a plutonium-based bomb was chosen for the test, which was named THE TRINITY. This name, which has reference to the Hindu divine Trinity, Brahma, Vishnu and Shiva, was believed to have been chosen by Oppenheimer, who was an avid reader of Sanskrit literature. A hundred foot-high steel tower with a platform was built for exploding the device with an estimated energy release 15,000 TNT equivalent. The plutonium bomb was set off on the morning of July 16, 1945. Many of the scientists led by Robert Oppenheimer were located in an observation shelter about 10,000 yards from the explosion site from where they could observe the explosion. There was lightning, for a brief period, of brightness equal to several suns at midday. A considerable part of the steel from the tower vapourised. The sound was heard even at about 100 miles. To quote the physicist I.I. Rabi, "There was an enormous ball of fire which grew and grew and it rolled as it grew; it went up in the air in yellow flashes and into scarlet and green…" Several thousand tonnes of dust and huge quantities of radioactive material formed during the fission created on the ground went up in the air and a crater of about 1,200 feet diameter was formed. No casualties or loss of property was reported. Watching the dramatic blast, Robert Oppenheimer remembered the following line from the Bhagavadgita, "I am the death, the destroyer of the worlds." General Grover in his report dated July 18, 1945, said, "I no longer consider the Pentagon a safe shelter from such a blast."

1.8 Use of Atom Bombs in Warfare

After the Trinity Test, the media reports declared in no uncertain terms that America had come to possess an extraordinarily destructive weapon. With the surrender of Germany, the war in Europe came to an end. Japan was also reported to be sending out peace feelers. President Truman was however advised by his military commanders to use the new weapon on Japan to bring a quick end to the war. Many scientists, who took part in the development of the nuclear weapon, including Leo Szilard, were however opposed to the use of this devastating weapon on moral grounds. But Truman went along with the advice of his military commanders.

On the morning of August 6, 1945, the first nuclear weapon "Little Boy" was dropped on the city of Hiroshima. The bomb exploded approximately 1800 feet above the ground with an explosive power equal to 12–15 kilotons of TNT, causing 100,000 deaths. Within a few days, on August 9, 1945, a second nuclear bomb "Fat Man" was dropped on the city of Nagasaki, causing approximately 45,000 deaths. The intensity was equivalent to 20–22 kilotons of TNT. Japan surrendered soon after and World War II ended.

1.9 Electricity Generation

Wells also forecast that the energy from a small quantity of uranium "when turned into a machinery could keep Edinburgh brightly lit for a week." In the course of the nuclear weapons programme, the Soviet Union and the West acquired considerable technological expertise for extraction of heat energy from nuclear fission in a controlled manner through compact long-lasting power sources (nuclear reactors) for direct use in shipping and submarines or for electricity generation. The first reactor to produce electricity was the small EBR-I reactor designed and developed by the US. Located in Idaho, this reactor began operating in December 1951. In the Soviet Union, an existing graphite-moderated plutonium-producing reactor located in Obninsk was modified for 5MWe (30MWt) power generation and became operational in June 1954. A demonstration Pressurized Water Reactor (PWR) with 60MWe power output became operational at Shippingport, the US in 1957. The fuel was enriched uranium with water as a moderator. With a monopoly over uranium enrichment, the US concentrated its efforts on PWR type reactors for power generation. The British built graphite-moderated natural uranium reactors of 50 MWe at Calder Hall. These reactors were designed for plutonium production as well as power generation. The first fully commercial reactor of the PWR type with 250MWe power output became operational in 1960 at Yankee Rowe in the US. France and Canada also began construction of nuclear reactors for power generation. The first two Soviet nuclear power plants were commissioned at Beloyarsk (Urals) and Novovoronezh (Volga region) in 1964. During this period, the main focus was on the expansion of the capacity of nuclear power plants. While initially, an average power plant could produce only a single gigawatt, within decades an average plant was producing 100 gigawatts.

The US President General Eisenhower's address entitled "The Atoms for Peace" before the UN General Assembly in December 1953 focused attention on the peaceful uses of atomic energy and the Geneva Conference (1955) gave fillip to nations to develop nuclear technology for the construction of nuclear power plants with technical assistance from the US and Soviet Union. The oil crisis of 1973 had a major effect on several countries, which relied heavily on oil for their energy needs. To meet this crisis, nations pushed for nuclear power plants as a substitute. For example, France was depending on oil for its energy supplies until 1974. By 2006, it had replaced nearly 80 percent of its power supply with nuclear power plants. Currently, 31 nations are engaged in nuclear power generation with over 400 nuclear reactors in operation. Another reason for the expansion of nuclear power was to mitigate the buildup of greenhouse gases in the atmosphere from fossil fuel burning thermal power plants. An MIT report (2009 update) says that currently, nuclear electricity accounts for over 16 percent (350 GWe) of the world market share. This was projected to increase to 20 percent (1000 GWe) by 2050.

To begin with, the nuclear power industry developed along many trajectories with power reactor developers in different countries introducing different types of nuclear technology. Eventually, the light water technology with enriched uranium as fuel first introduced in the US came to dominate the global nuclear power industry. Boiling Water Reactors (BWR) based on this technology now account for more than 90 percent of installed nuclear capacity worldwide. The other power reactor technology with an international presence is the Canadian CANDU design that uses natural uranium with heavy water as the moderator.

Simultaneously, the nuclear power industry has been developing and improving reactor technology that could be safer and more efficient through modifications in existing designs as well as incorporating radical changes. The creation of nuclear fuel cycles based on the new state-of-the-art reactor technologies will also facilitate multifold growth of energy efficiency through the use of the predominantly available uranium-238 and thorium-232 as fertile materials for breeding plutonium-239 and uranium-233 fissile materials. Research and industry are also addressing the environmental concerns about the nuclear power technologies including the safe disposal of spent nuclear fuel. In achieving these goals, the nuclear power industry has before it the following challenges:

1. Enhanced Safety and increased efficiency
2. Economic competitiveness
3. The country's nuclear energy strategy
4. Effective regulation and licensing process
5. Spent fuel and waste management
6. Proliferation control, and
7. Public acceptance

The current stage of nuclear power industry development marks the beginning of nuclear power plant evolution in the 21st century.

1.10 Nuclear Power Plant Accidents

Nuclear power generation has, however, been the subject of serious debate ever since the use of nuclear weapons. This debate intensified with two major nuclear reactor accidents during the 70s and 80s in the wake of the Three Mile accident in 1976 and the Chernobyl accident in 1986 resulting in a setback in the growth of nuclear power. Just when expansion activity began reviving at the beginning of the 21st century, a major tsunami disabled three power reactors in Fukushima Daiichi in March 2011 and it sparked revived opposition to nuclear power generation.

In spite of these accidents, the overall momentum for the modernisation and expansion of nuclear power generation remains steady.

1.11 Uranium, the Only Natural Element for Nuclear Power Generation

With the ultimate depletion of non-renewable fossil fuel resources, nuclear power generation is currently, the only major option to meet the global energy needs of the 21st century. It is also the major option to reduce the greenhouse gases from fossil fuels which are already affecting the climate system.

Uranium is the only natural element that plays a pivotal role in nuclear reactor technology. An attempt will be made to present the reader with a general account of the use of this precious element for energy generation with a special emphasis on India's nuclear power programmes.

The potency of uranium dreamed of by H.G. Wells in 1914 has thus proved to be a reality a few decades later.

2
Uranium and Radioactivity

2.1 Discovery of Radioactivity

The end of the 19th century saw the discovery of a new phenomenon called "radioactivity." Antoine Henry Becquerel (1852–1908) discovered it in the course of his studies on the fluorescence and phosphorescence exhibited by some chemicals exposed to light and X-rays. Fluorescence represents the emission of radiation especially of visible light, stimulated in a substance by absorption of radiation and persisting only as long as the stimulating radiation is continued. Phosphorescence represents a persistent emission of light following exposure to and removal of incident radiation. Becquerel, born in a family of scientists, inherited not only his interest in this subject but also some fluorescing compounds (including uranium compounds) used in these investigations from his father. His studies received fresh impetus in 1895 when Wilhelm Conrad Röntgen discovered X-rays from the luminescence they produced on a sheet of paper coated with barium platinocyanide. Röntgen also observed that the X-rays caused darkening of a photographic plate even when it was wrapped in a black paper. Becquerel decided to establish a connection between the X-rays and the fluorescence exhibited by the uranium salts on exposure to sunlight. He wrapped a photographic plate in black paper, placed a thin crystal of the uranium salt upon the paper and then exposed the setup to sunlight. On developing, the plate displayed dark images of the crystals. He concluded that the uranium salt, on exposure to the sun, emitted X-rays that could penetrate the opaque paper and blacken the photographic plate.

An unexpected event led to his discovery of path-breaking concepts on the structure of matter. During the last week of February 1896, the sky over Paris was overcast for a few days with the sun shining only intermittently. Becquerel left the ensemble of uranium crystals together with the covered photographic plate in a drawer for a few days in a drawer planning to proceed with the experiment when sunshine came back. But providentially, he decided to develop the photographic plate without exposing it to sunlight and expecting it to find no darkening of the plate. To his amazement, the supposedly unexposed plate showed clear and strong images of the crystals. He then concluded that the crystals emitted penetrating radiation even without exposure to sunlight and that this radiation persisted for a long time. What he had actually discovered was a remarkable phenomenon, which Marie Sklodowska named 'radioactivity' in 1898. Becquerel also reported that the radiation emitted by uranium had some properties different from X-rays. They could be deflected by a magnetic field and therefore, he hypothesised they must consist of charged particles. He also found that all uranium-containing materials gave off spontaneous emissions while none of the fluorescent compounds, other than uranium, did. Becquerel and the Curies were awarded the Nobel Prize for Physics in 1903 for their classic discoveries in radioactivity.

2.2 Discovery of Radium

Making use of the ionisation chamber developed by her husband, Marie Curie noted that two of the uranium-containing minerals, pitchblende, and torbernite, were much more radioactive than the uranium they contained. She suspected that the minerals contained traces of an unknown element, which is more radioactive than uranium. Starting with several tonnes of Joachimstahl pitchblende residues left after extraction of uranium, which were at her disposal thanks to the good offices of the Austrian Academy of Sciences, Marie Curie set out to separate these active species employing elaborate chemical procedures. Finally, she succeeded in separating and identifying two new elements, – one similar to bismuth, and another similar to barium. The first element was named 'Polonium' after the native land (Poland) of Marie Curie and the other 'Radium,' meaning 'ray' in Latin. Both elements were million times more radioactive than the same amount of uranium. The French chemist André-Louis Debierne, an associate of the Curies, isolated yet another element 'Actinium' from pitchblende. Pierre died in 1906 in a road accident. The 1910 Radiological Congress honoured him by naming the radioactivity associated with one gram of radium, the basic unit of radioactivity, as a Curie (Ci). A year later in 1911, Marie was awarded the Nobel Prize for Chemistry for her discovery of polonium and radium. For the rest of her life, Marie worked tirelessly promoting the use of radium for the treatment of cancer. Those were the days when very little was known about the harmful effects of radiation. Mme. Curie died on July 4, 1934, from pernicious anemia, most probably caused by severe radiation exposure.

2.3 Foundations of Nuclear Science

The late 19th century discoveries that led to the demonstration of the subatomic nature of radioactivity and the isolation of two highly radioactive elements, polonium and radium, by the Curies necessitated the revision of centuries-old ideas about the immutability of the atom. This led to the emergence of the new discipline of nuclear science. These discoveries will now be described briefly.

2.3.1 Discovery of the Electron

A few years before Becquerel discovered radioactivity, H. Geissler, W. Hittorf, E. Goldstein, W. Crookes, and J. Perrin studied the electrical discharge through gases contained in tubes at low pressure. The period also saw fundamental investigations on the phenomenon of electrolysis by Michael Faraday. And J.J. Thomson, in England, who carried out more comprehensive studies, established the nature of the discharge (called the cathode rays) emitted at the cathode. From the bending nature of these rays under the influence of an electric field, Thomson concluded that they consisted of a stream of negatively charged particles. They were later renamed 'electrons.' All these studies prompted Thomson to question the fundamental hypothesis of Dalton's Atomic Theory that atoms are indivisible. In 1897, he postulated that atoms consist of 'corpuscles' (electrons) floating in a sea of positive charge. This is called Thomson's 'Plum Pudding Model' of the atom. For these epochal studies, Thomson won the Nobel Prize for Physics in 1906.

2.3.2 Nature of Radiation Emitted by Uranium

Becquerel, the Curies, and Ernest Rutherford and colleagues played an important role in establishing the nature of the radiation emitted by uranium.

Alpha, Beta and Gamma Rays: Mme. Curie found that when a narrow beam of radiation emitted by uranium minerals was passed between electrically-charged plates, it separated into three parts: one part deflected towards the negative electrode, indicating that it contained positively-charged radiation; a second part deflected towards the positive electrode, showing its negative charge; and a third passed without deflection, showing its neutral nature (Fig.1). The positive rays were called 'alpha' (α) rays, the negative rays 'beta' (β) rays and the neutral ones 'gamma' (γ) rays.

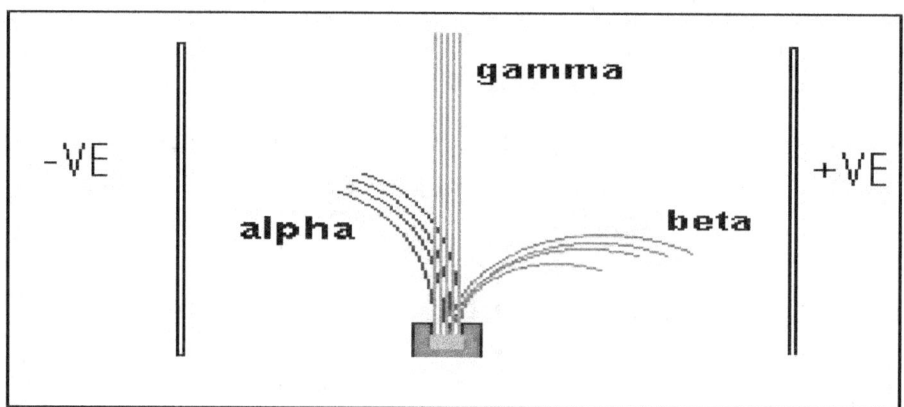

Fig. 1: Resolution of alpha, beta and gamma Radiations in a Magnetic Field

On the basis of their charge and charge/mass ratio, the negatively-charged beta rays were shown to consist of electron particles. Following the observation of William Ramsay and Frederick Soddy that helium occurred in appreciable quantities in minerals containing uranium and thorium, Rutherford and his team collected the alpha particles from a radioactive source in an evacuated glass tube. The spectrum of an electric discharge through the gas was found to be identical to that of helium. They established that the alpha particles are positively-charged helium atoms (helium ions). Villard, the French chemist, showed that the fraction of radiation, which is not deflected by an electric or magnetic field, was akin to X-rays (electromagnetic radiation) but of a shorter wavelength. These radiations were named gamma (γ) rays.

Penetrating Power of Radiations: Rutherford and his team found that the penetrating power of the three types of radiation differed considerably. While the alpha rays could be stopped by a few tens of centimeters of air or a thin sheet of paper, beta rays were 100 times more penetrative and could pass through several meters of air or a thin aluminum sheet. Gamma rays were highly penetrating and would require thick concrete or a lead sheet a few centimeters thick to block them.

Scattering of α-particles: In 1909, Rutherford and his colleagues, Hans Geiger and Ernst Marsden, studied the scattering of α-particles from polonium by a thin gold foil. While most of the particles passed through the foil with minor deflection from their straight path, a small number of particles were deflected through large angles, some even rebounded through the path of incidence (Fig.2).

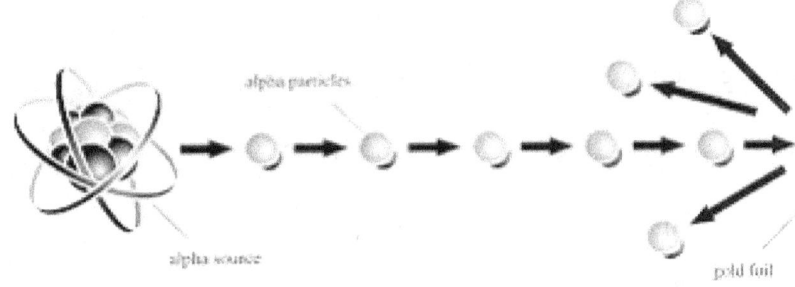

Fig. 2: Scattering of Alpha Particles by a thin Gold Foil

From this observation, it was concluded that most of the atom of 10^{-10} cm radius is empty with almost the whole of the mass of the atom and its positive charge concentrated in a relatively small volume (10^{-15} to 10^{-14} cm) of the atom which was named the *nucleus*. To account for the fact that they did not fall into the positively-charged nucleus, the negatively-charged electrons are postulated to be in a state of rapid rotation around the nucleus at some distance, much like the planets revolving round the Sun. The atom as a whole is neutral with equal positive and negative charges. This is called the Rutherford Model of the Atom. There was, however, a lacuna in this model. According to classical electromagnetic theory, a charged particle moving in a curved path loses energy by emitting electromagnetic radiation and collapses into the nucleus. Niels Bohr in 1913 refined this model and proposed a quantised shell model in which the electrons outside the nucleus move in orbits of fixed size and energy (stable orbits). They absorb or emit energy only when they jump from one orbit to another. Heisenberg and others introduced elaborate wave mechanical modifications to explain the behaviour of multi-electron systems.

2.4 Radioactive Decay

In 1900, William Crookes observed that on adding ammonium carbonate to a solution of uranium nitrate a precipitate of uranium carbonate was first formed. On the addition of excess reagent, this precipitate almost entirely dissolved leaving behind a small quantity of the precipitate in which the entire radioactivity was concentrated; while the solution actually containing all the uranium was almost inactive. He named the active portion 'Uranium-X.' In 1902, Rutherford and Frederick Soddy separated from a solution of thorium nitrate also a fraction which contained almost all the radioactivity, which they named 'Thorium-X.' These workers further observed that in both cases the activities in the active fractions (UX and ThX) decreased at characteristic rates, while there was a simultaneous growth in the activities at corresponding rates in the inactive portions. They further found that these decay and growth processes were independent of temperature, pressure and other factors that normally affect the rates of chemical reactions. From these observations, Rutherford and Soddy concluded that radioactivity is a nuclear property. By following the change in activities with time, they formulated the exponential decay law (depicted below).

$$N_t = N_0 e^{-\lambda t}$$

where

N_0 – number of atoms present at any arbitrary time

N_t – number of atoms remaining after lapse of further time interval t

λ – radioactive decay constant

The radioactive decay constant λ (or decay constant) is a definite and specific property of a given radioactive species. Its value depends only on the nature of the elemental species, and is independent of its physical condition or state of chemical combination. In simple words, the law states that a fixed fraction of the element (say, half of it) decays in a fixed time. The time required for this half decay is called the 'half-life $(t_{1/2})$' of the radioactive element and is given by the relationship:

$$t_{1/2} = 0.693/\lambda$$

If half the radioactive atoms decay in four days ($t_{1/2}$ = 4 days), half of the remaining atoms decay in the next four days and so on. The half-life period is characteristic of a radioactive element. The half-life values of radioactive species may range from a fraction of a second (e.g. Bi-209 = 10^{-23} sec) to a few billion years (e.g. U-238 = 4.5 billion y).

For his investigations into the disintegration of the elements and the chemistry of radioactive substances, Rutherford was awarded the Nobel Prize in Chemistry in 1908. While receiving the award, Rutherford quipped, "I have dealt with many transformations with various periods of time, but the quickest that I have met was my own transformation in one moment, from a physicist to a chemist."

2.4.1 Displacement Law, Isotopes and Radioactive Decay Series

Following the isolation of UX and ThX, several more radioactive species were identified. While some species could be distinguished as separate elements, there were several species that were chemically indistinguishable. This complexity was resolved in 1902 by Rutherford and Soddy, who observed that a radioactive element is unstable and undergoes spontaneous transformation by the emission of α – or β – particle from its nucleus. The resulting atom has chemical and physical properties that are different from the parent atom. This atom, in turn, may be unstable and pass through a succession of transmutations. Soddy formulated a law known as the 'Displacement Law' for these transmutations. In terms of this law, a radioactive atom 'A' emitting an α-particle, followed by two successive β-particles gives rise to atom 'B' that is the chemically same as atom 'A' but with a reduced mass of four units. Soddy named these atoms, Isotopes; they are from the same element and have different atomic masses. The 30-odd radioactive species that were isolated were identified as isotopes arising from radioactive transmutation of the ten radioactive elements beyond lead in the Periodic Table. It was further observed that these species fall into three distinct decay series for which the parents are U-238, U-235, and Th-232. Soddy was awarded the Nobel Prize in Chemistry in 1922 for his contribution.

In 1920, Francis William Aston, using his 'mass-spectrograph' showed that most of the naturally occurring elements, radioactive or stable, are also made up of isotopes, with only 20 elements (e.g. fluorine) that were exceptions. There are 280 isotopic forms of stable elements and about 40 radioactive isotopes occurring naturally. In addition, more than 1200 radioactive isotopes have been prepared synthetically. Aston was awarded the Nobel Prize in Chemistry in 1922 for his contributions.

2.5   Discovery of the Neutron

In 1920, Rutherford predicted the possible existence of a neutral particle as a constituent of the atom. According to him, a proton in combination with an electron might form a neutral doublet that moves freely through matter since it carries no charge. This hypothetical particle was named 'neutron' by W. D. Harkins in 1921. Three years later, James Chadwick identified this particle in the reaction between alpha particles and beryllium. Chadwick was awarded the Nobel Prize in Physics in 1935 for this discovery.

$$_4Be^9 + _2He^4 \rightarrow _6C^{12} + _0n^1$$

With the discovery of the neutron, it was possible to explain all nuclear phenomena on the hypothesis that protons and neutrons constitute the building blocks of the nucleus of atoms. A nucleus consists of Z number of protons (atomic number) and N number of neutrons with a mass number A= (N+Z). While the protons and neutrons are of nearly the same mass, the protons are positively charged and neutrons have no charge. They are represented as $_1H^1$ (proton), $_0n^1$ (neutron), respectively. The number of electrons in the extra-nuclear orbits is the same as the number of protons in the nucleus (or the atomic number Z). Thus, the atom as a whole is electrically neutral. Electrons with very low mass value do not make a significant contribution to the mass of the atom. The atomic number Z uniquely identifies an element. The isotopes of an element consist of the same number of protons but differ in the neutron number in the nucleus (e.g. U^{238} and U^{235}). A convention is followed for designating the atomic number (Z) and mass number (A) of an atom – $_ZX^A$ (X=atom).

How are beta particles (electrons) emitted from the nucleus if it carries only protons and neutrons? An explanation for this was provided by Wolfgang Pauli in 1931, according to which the beta particle emission occurs through a neutron transforming into a proton, an electron, and a neutrino. The neutrino carries negligible mass and no charge. The electron and neutrino are ejected from the nucleus while the proton remains in the nucleus. The existence of the neutrino was experimentally established 20 years later in 1951.

2.6 Radioactivity of Uranium

In the light of facts mentioned above, the radioactivity of natural uranium can now be explained.

Uranium: Atomic number – 92

Atomic weight of natural uranium – 238.029

Isotopes of uranium – 238, 235, 234 (Mass numbers)

Atomic composition: (P = Proton, N = Neutron)

U-238 Nucleus contains 92 P + 146 N and 92 outer electrons

U-235 Nucleus contains 92 P + 143 N and 92 outer electrons

U-234: Nucleus contains 92 P + 142 N and 92 outer electrons

Isotopic composition (per cent by weight): U-238 = 99.27; U-235 = 0.72; U-234 = 0.0055

All the three isotopes are radioactive. Their half-lives are (approx.):

U-238 — 4.5 billion (4,500 million) years

U-235 — 700 million years

U-234 — 25,000 (0.025 million) years

The half-life is also a measure of the number of atoms that disintegrate per second; the shorter the half-life, the larger the number of disintegrations per unit time. For a given weight, U-234 undergoes the largest number of disintegrations per unit time, followed by U-235 and U-238 (the lowest).

The nuclei of uranium isotopes are unstable and undergo radioactive disintegration through a series of steps till they reach a stable state like lead atoms. Uranium-238 and 235 undergo a series of α and β decays eventually leading to the formation of the stable isotopes of lead-206 and 207, respectively (Figs. 3 & 4). In a similar manner, thorium-232 undergoes decay producing the stable isotope, lead – 208.

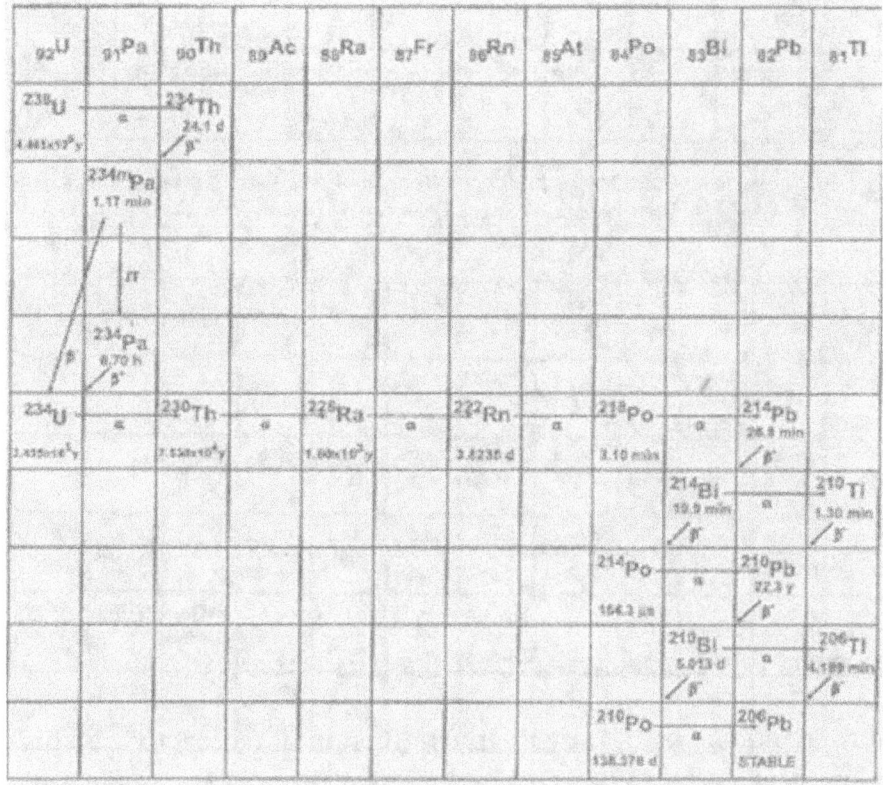

Fig. 3: Uranium-238 Decay Series

2.7 Units of Radioactivity

The radioactivity associated with one gram of radium-226 was defined as the Curie (Ci). One gram of radium-226 undergoes 3.7×10^{10} disintegrations per second. The Curie is now defined as that quantity of a radioactive substance which decays at the rate of 3.7×10^{10} disintegrations per second (dps). Weight for weight, a material with long half-life carries much less activity compared to a material with a shorter half-life. For example, 1 kg of uranium-238, with a half-life of 4.5 billion years carries only 0.00033 Ci of radioactivity, while the same amount of cobalt-60, with a half-life of 5.3 years carries as much as one million Curies. The radioactivity associated with unit mass (say 1 gram) of a substance is called its 'specific radioactivity.'

In the SI system, the unit of radioactivity is the Becquerel (Bq) with a value of one disintegration per second. Thus,

$$1 \text{ Ci} = 3.7 \times 10^{10} \text{ Bq}$$

and

$$37 \text{ billion Bq} = 37 \text{ G Bq} = 1 \text{ Ci}$$

$$1 \text{ billion Bq} = 1 \text{ G Bq} = \text{about } 27 \text{ mCi}$$

$$1 \text{ million Bq} = 1 \text{ M Bq} = \text{about } 27 \text{ micro Ci}$$

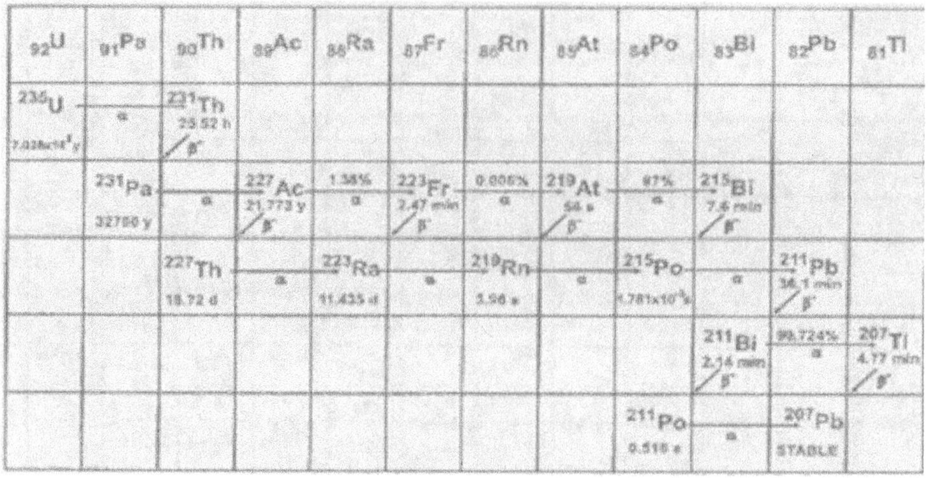

$_{92}$U	$_{91}$Pa	$_{90}$Th	$_{89}$Ac	$_{88}$Ra	$_{87}$Fr	$_{86}$Rn	$_{85}$At	$_{84}$Po	$_{83}$Bi	$_{82}$Pb	$_{81}$Tl
^{235}U — α →		^{231}Th 25.52 h — β⁻ →									
	^{231}Pa 32760 y — α →		^{227}Ac 21.773 y — 1.38% α / β⁻ →		^{223}Fr 2.47 min — 0.006% α / β⁻ →		^{219}At 54 s — 97% α / β⁻ →		^{215}Bi 7.6 min — β⁻ →		
			^{227}Th 18.72 d — α →	^{223}Ra 11.435 d — α →		^{219}Rn 3.96 s — α →		^{215}Po 1.781×10⁻³ s — α →		^{211}Pb 36.1 min — β⁻ →	
									^{211}Bi 2.14 min — 99.724% α / β⁻ →		^{207}Tl 4.77 min — β⁻ →
								^{211}Po 0.516 s — α →		^{207}Pb STABLE	

Fig. 4: Uranium-235 Disintegration Series

2.8 Detection and Measurement of Radioactivity

Becquerel used a photographic plate to study the phenomenon of radioactivity. This was later substituted with a gold leaf electroscope because of its higher sensitivity. This was followed by the ionisation chamber developed by the Curies. It consisted of two horizontal plates connected to a high voltage battery and a sensitive galvanometer. The radioactive sample was placed on the bottom plate. The radiation caused ionisation of the air between the plates and this, in turn, caused an electric discharge which was detected by the galvanometer. The magnitude of the discharge, which was a measure of the ionising power of the radiation from the source, was then measured by balancing it with the electricity produced by applying pressure to a piezoelectric quartz crystal. All these methods could be used only to measure 'bulk' radiation and were capable of indicating individual radiation events.

The spinthariscope, developed by Crookes in 1903, was the first instrument that could detect individual radioactive disintegration events. It consisted of a small screen coated with luminescent zinc sulphide. When exposed to alpha particles the scintillations created by the individual particle collisions with the screen were shown. These collision events could be counted using a magnifying glass.

An improved design of this instrument was used by Rutherford and his coworkers in their path-breaking studies that led to the identification of the atomic nucleus. However, the method proved very cumbersome causing strain and fatigue in the observers.

Another instrument that was used in the early years was the 'Cloud Chamber' developed in 1911 by C. T. R. Wilson. This consisted of a sealed chamber containing air saturated with water vapour. The passage of a charged particle through the chamber condenses the vapour in its path into tiny droplets. A visible trail of these droplets gives a picture of the passing radiation. Alpha particles leave a broad, straight path of definite length. The beta particles display a bent trail while the gamma rays leave no visible trail because of their poor ionising power. While the Cloud Chamber helped the discovery of some nuclear particles (e.g. the positron), it was not very helpful for routine measurements. These devices were superseded by more sensitive and specific instruments.

2.8.1 Geiger – Müller Detector

The Geiger-Müller (GM) detector is the best known among the radiation counting instruments based on the ionising effects of nuclear radiations. Invented by Hans Geiger and Wilhelm Müller in 1928, the device counts the particles emitted in radioactive decay as pulses of electric current.

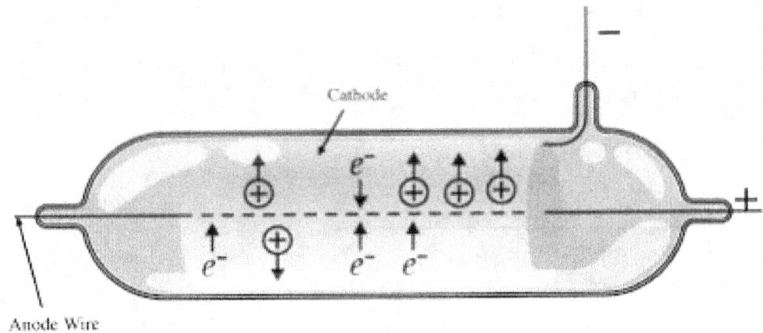

Fig. 5: Geiger-Muller Detector

The GM detector consists of an evacuated glass tube filled with argon and a small amount of alcohol vapour at low pressure. A wire stretching along the axis of the tube is maintained at a high positive voltage with respect to the surface of the tube. When an alpha or beta particle enters the tube through a thin window, it ionises an atom of the gas; i.e., it knocks off an electron from an atom of the gas in the tube. The free electron accelerating towards the central wire at a high voltage knocks off more electrons from the gas molecules. This results in an electron 'avalanche,' which produces an easily detectable electric pulse that can be recorded electronically. The GM detector, because of its simplicity, ease of operation and cheapness, continues to be the best choice for routine counting operations. Its normal efficiency for gamma-ray counting is, however, low.

A GM detector per se cannot detect neutrons because they do not cause ionisation. Their detection and measurement are, therefore, based on the secondary ionisation produced by their interaction with suitable elements. For example, the detectors for slow neutrons are based on the detection of the alpha particles produced in the (n, α) reaction with B^{10} or Li^6. A widely used neutron detector is an ionisation tube filled with boron trifluoride gas.

2.8.2 Scintillation Detectors

The scintillation method (spinthariscope) used by Rutherford and colleagues came into vogue again in the 1940s with the introduction of photomultiplier tubes that convert the scintillation events and enable electronic counting. The counting system consists of a scintillation detector system with a photomultiplier (Fig.6) and a signal processor and recorder. Of the several inorganic scintillating media that meet the required criteria, the thallium-activated sodium iodide crystal has become the most widely used material for γ-ray detection (in γ-ray spectrometer). Caesium iodide activated with sodium is another detector. Solid-state semiconductors such as Germanium or Silicon doped with Lithium [Ge(Li), Si(Li)] and High Purity Germanium (HPGe) are also extensively used for γ-ray measurements. Being 1,000 times denser than gases their dimensions can be kept much smaller. Organic scintillation detectors are used mainly for alpha and beta measurements. Simple organic scintillation materials include naphthalene, anthracene, and stilbene. An example of a liquid organic scintillation detector is PPO (2, 5-diphenyoxazole) with a wavelength shifter POPOP [1, 4-(5-phenyloxazolyl) benzene]. Plastic scintillators are made by mixing liquid scintillators (e.g. tetraphenyl butadiene) and a monomer such as styrene followed by polymerisation and moulding into the required shape.

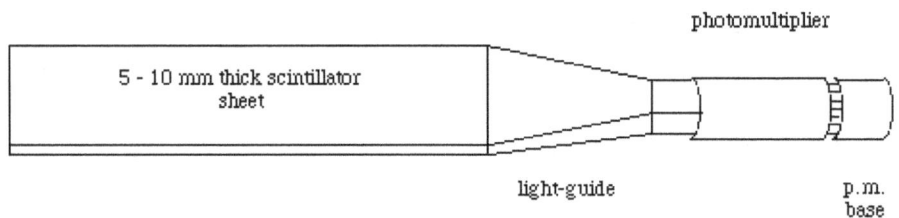

Fig. 6: Scintillation Detector

2.9 Applications of Radioactivity Counting Systems

The rapid advances in radiation detection and measurement techniques are entirely due to the phenomenal developments in electronic devices for receiving and processing the signals from radiation detectors. With advances in semiconductor technology, compact, reliable and low energy counting devices with a wide range of counting rates have become available to meet specific needs. The application of computer techniques is yet another landmark in radiation measurements. Thanks to these advances, a wide range of radioactivity measuring instruments for use in radioactive mineral exploration, nuclear engineering, medicine, agriculture, hydrology, and industry. They also help monitor environmental hazards from radiations and take prompt corrective steps.

In the matter of handling of uranium, the use of radiation detectors starts with exploration for minerals and extends to the complete chain of operations involving processing and all the subsequent stages, including reactor operation and control. Continuous monitoring of radioactivity is a mandatory requirement in the entire nuclear power industry.

3

Uranium and Nuclear Fission

3.1 Artificial Nuclear Transmutation

The observation that the natural transformation of radioactive elements occurs through the emission of an alpha or a beta particle led scientists to speculate if the nuclei of normally stable elements could be transmuted by bombarding them with energetic particles. Such a possibility was established when Rutherford bombarded nitrogen, contained in a chamber, with alpha particles from Polonium-210 and identified the product as a new isotope of oxygen with a mass of 17. This reaction, announced in 1919, was the first instance of 'artificial nuclear transmutation.' The reaction, called an (α,p) reaction, can be represented as,

$$_7N^{14} + {}_2He^4 \rightarrow {}_8O^{17} + {}_1H^1$$

(α-particle) (proton)

Similar reactions were carried out by Rutherford and others using different target elements like $_{11}Na^{23}$, $_{13}Al^{27}$, $_{16}S^{32}$, $_{19}K^{39}$. But it was not possible to use heavier targets beyond potassium as the energy associated with the positively-charged alpha particles from Polonium-210 was not sufficient to get over the Coulomb repulsion between the positively charged alpha particles and the positively-charged heavier target nuclei. However, using particle accelerators such as the cyclotron the alpha particles and other positively-charged projectile particles protons and deuterons could be accelerated to the energy levels required to get over the Coulomb repulsion and bring about several nuclear transformations.

When the news of nuclear transmutation reactions was announced, The New York Times commented, *"Science has obtained conclusive proof from recent experiments that innermost citadel of matter, the nucleus of the atom, can be smashed, yielding tremendous amount of energy and probably vast new stores of gold, radium and other valuable minerals."* In contrast, Rutherford, exemplifying the sobriety of a scientist, said, *"The transformation of the atoms is of extraordinary interest to scientists but we can not control atomic energy to an extent which would be of any value commercially, and I believe we are not likely ever to be able to do so."*

3.2 Discovery of the Neutron

While experimenting with alpha particles, Walter Bothe, a German physicist, observed in 1928 that when beryllium was bombarded with alpha particles from polonium, a penetrating, electrically neutral radiation was emitted. He mistook this for gamma radiation. In 1932, James Chadwick worked with Rutherford and repeated the Bothe's experiment. They showed that the radiation consisted of neutral particles with a mass approximately equal to that of a proton. This particle identified as the neutron was the neutral particle earlier postulated by Rutherford. The reaction is represented as,

$$_4Be^9 + {}_2He^4 \rightarrow {}_6C^{13} \rightarrow {}_6C^{12} + {}_0n^1$$

(neutron)

This type of reaction is called an (α, n) reaction. Because it carries no charge, a neutron is not repelled by the target nucleus. For this reason, neutrons do not need to be accelerated to high energies to bring about nuclear reactions. On the other hand, slow neutrons are usually more effective than fast ones, because fast neutrons may pass through the target nucleus without interaction. Nuclear reactions involving neutrons are also easier and cheaper to perform than those requiring positively-charged particles such as alpha particles. Neutron reactions can often be carried out in a small laboratory using a neutron source. An (α,n) source consisting of an α-emitter such as radium mixed with beryllium is found to be very convenient for this purpose.

3.3 Artificial Radioactivity

In 1934, French chemists Irène Joliot-Curie and Frédéric Joliot (daughter and son-in-law to Marie Curie), bombarded an aluminium foil target with α–particles and found that the target exhibited activity which persisted even after the bombardment stopped. Using elegant separation methods, they were able to establish that the activity arose from the artificial radioactive product 30P, which decayed with a characteristic half-life of about three minutes through positron emission.

$$_{13}Al^{27} + {}_2He^4 \rightarrow {}_{15}P^{30} + {}_0n^1 - (\alpha, n) \text{ reaction}$$

$$_{15}P^{30} \rightarrow {}_{14}Si^{30} + {}_1e^0$$

(positron)

During the same period Enrico Fermi, an Italian physicist and his co-workers procured a strong radium-beryllium neutron source and began bombarding several elements with the neutral projectiles. They observed multiple (n,γ), (n,p) and (n,α) reactions yielding beta-active radioactive products. Examples are:

$$_{49}In^{115} + {}_0n^1 \rightarrow {}_{49}In^{116} + \gamma - (n,\gamma) \text{ reaction}$$

$$_{49}In^{116} \rightarrow {}_{50}Sn^{116} + {}_{-1}e^0 \ (t_{\frac{1}{2}} - 13 \text{ sec})$$

$$_7N^{14} + {}_0n^1 \rightarrow {}_6C^{14} + {}_1H^1 - (n, p) \text{ reaction}$$

$$_6C^{14} \rightarrow {}_7N^{14} + {}_{-1}e^0 \; (t_{1/2} - 5568 \text{ y})$$

$$_{13}Al^{27} + {}_0n^1 \rightarrow {}_{11}Na^{24} + {}_2He^4 - (n, \alpha) \text{ reaction}$$

$$_{11}Na^{24} \rightarrow {}_{12}Mg^{24} + {}_{-1}e^0 \; (t_{1/2} - 15 \text{ h})$$

The following years saw the preparation of a large number of artificial radioactive isotopes, which contributed immensely in chemical and biological investigations.

3.4 Neutron Irradiation of Heavy Elements

In 1934, Fermi and his group extended their experiments with neutrons all the way up to uranium, the heaviest naturally-occurring element with atomic number 92. They found that uranium too gave rise to new beta active species with very short half-lives. The team surmised that the initial product (239U) emitted a beta particle and transformed into the next higher elements beyond uranium (At Nos. 93 and 94). They collectively called these elements *transuranium* elements or *transuranics*. Their properties were expected to be similar to rhodium, osmium, iridium, and platinum (the platinum group metals). But this conclusion could not be confirmed experimentally. Ida Noddack, a German chemist, who was working with her husband, suggested that Fermi compare the chemistry of their *transuranium* elements, not only with those in the neighbourhood of uranium but with all known elements. She said, *"It is conceivable that the nucleus breaks up into several large fragments, which would, of course, be isotopes of known elements but would not be the neighbours of the irradiated element."* In making this suggestion, she foretold the possibility of what later came to be known as nuclear fission. Ironically, Noddack herself did not undertake experimental verification, nor did the suggestion attract the attention of nuclear scientists.

3.4.2 Discovery of Fission

Otto Hahn, a German chemist, after getting his doctorate in Germany in 1901, went to England and worked with Sir William Ramsay. He then moved to Montreal, Canada to work with Rutherford on radioactivity. After gaining experience in the field, he returned to Germany in 1906 as the Head of Radiochemistry Department at the famous Kaiser Wilhelm Institute (later called the Max Plank Institute). Working with Lise Meitner, a nuclear physicist, he isolated a new element, protactinium-231. In 1934, Hahn became keenly interested in Fermi's work on neutron irradiation of uranium and wanted to verify Fermi's assumption that elements beyond uranium (transuranics) are formed in this reaction. In July 1938, Meitner, a Jew by birth, fled to Sweden from Germany to escape persecution by the Nazis. Hahn continued his investigations with Strassmann as his associate.

In Paris Irène Joliot-Curie and P. Savitch also became interested in Fermi's work. They studied one of the beta active products of irradiation with a half-life of 3.5 hours. They first took it for an isotope of thorium, later for actinium and finally, for a transuranic element resembling lanthanum. Actually, their substance was lanthanum itself. But they were misled by some problems in their experimental work. Had they recognised this, they would have been on the right path of the discovery of nuclear fission!

Early in December 1938, Hahn and Strassmann were on the verge of agreeing with the findings of Joliot-Curies that actinium and thorium are among the products of neutron irradiation of uranium. These product elements were supposed to have formed by beta emission from the radium isotope, which they thought was the initial product of the nuclear reaction. For uranium (element 92) to undergo conversion to radium (element 88) two α-particles have to be emitted. But there was no experimental evidence for α-emission. Well versed in the art of chemical identification of radioactive elements in minute quantities Hahn and his associate attempted to establish the presence of radium, directly by radiochemical means. They dissolved the irradiated uranium target in acid and carried out carbonate precipitation after adding barium as a carrier for radium. They converted the purified mixed carbonate into bromide. Fractional crystallisation of the salt did not separate the supposed 'radium activity' from the 'carrier' barium. Other laborious fractional crystallisation procedures involving chloride and chromate also proved futile. In 1939, they reported as follows: *"As chemists, we must actually say that the new particles do not behave like radium but, in fact, like barium, but radioactive; as nuclear physicists, we can not make this conclusion, which is in conflict with all experience in nuclear physics."* In spite of strong chemical evidence from their results, and reluctant to go against the ideas of respected nuclear physicists of the day, they over-cautiously spoke of the "bursting" of uranium yielding barium (At Wt.137) and another element fragment (At Wt.101). On one occasion, Hahn is reported to have said, *"We are afraid of the physics people!"*

3.4.3 Confirmation of Nuclear Fission

Hoping to get his doubts cleared, Hahn wrote to Meitner, who was equally baffled by the results. It was not easy to explain how the entry of a neutron into the nucleus of uranium could cause such an upheaval, chipping off 100 particles, leading to the formation of barium with Z=56 and A=140 (nearly). At the same time, she had great confidence in Hahn's experimental skill. She posed this problem to Otto R. Frisch, her nephew, and a nuclear physicist, who working in Niels Bohr's Institute of Theoretical Physics in Denmark. After pondering the problem for a few days, Meitner and Frisch came out with their historical publication showing how Bohr's liquid drop model of the atom could explain the cleavage of a heavy nucleus into two nuclei of medium size. According to Bohr's theory, a heavy nucleus behaves like a liquid drop and not like a solid ball. A liquid drop when hit hard, might stretch, vibrate and finally break into two. The different stages in the process can roughly be shown as,

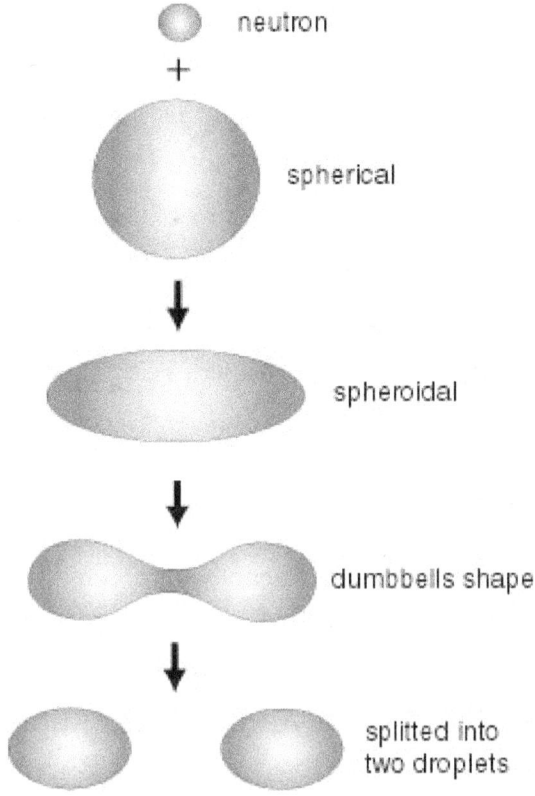

Fig. 1: Liquid drop Model of Nuclear Fission

Top of Form

Meitner and Frisch concluded that a similar process occurred when a neutron interacted with the uranium nucleus. They named the process "fission" and calculated that the energy release from fission amounted to 200 million electron volts. Subtracting the nuclear charge of barium (56) from that of uranium (92), they identified the other fission product was element 36, the inert gas krypton. Hahn and Strassmann later detected this element in the form of its decay product strontium. This was followed by the discovery of other fission products.

Meanwhile, Frisch proceeded to obtain experimental evidence for his hypothesis of energy release in the fission process. He constructed an ionisation chamber with an amplifier and oscilloscope. The invisible particles, formed when uranium was bombarded with neutrons passed through the ionisation chamber, would be detected as pulses on the oscilloscope. The size of the energetic pulse from the fission particles indicates the energy of the particle. Working overnight, in the wee hours of the morning in January 1939, Frisch found evidence for the energetic pulses. Calculations showed that the mass of the two daughter nuclei together would be less than that of the original uranium nucleus. This mass deficiency was equivalent to 200 MeV. This was also the first experimental confirmation of Albert Einstein's law of mass-energy equivalence published in 1905. Frisch wrote a paper presenting his experimental confirmation of uranium fission. Niels Bohr carried the news of the work on fission to the US.

These developments in 1939 sparked activity in many laboratories. Hahn and Strassmann conjectured that fission not only released a lot of energy but also released additional neutrons. This was later confirmed by Joliot-Curie, H. von Halban, L. Kowarski, and others. The liberation of additional neutrons made it possible to harness the vast amounts of energy released in the process of uranium fission through a chain reaction. Bohr proposed that fission was more likely to occur in uranium-235 than in uranium – 238, and more effectively with slow neutrons than with fast neutrons.

For his work on fission, Otto Hahn received the Nobel Prize in 1945.

3.5 Nuclear Fission and Energy Release

Energy release from the fission of a uranium nucleus can be explained in terms of *nuclear mass* and *binding energy*. The magnitude of the rest masses of the building blocks of an atom are so small that they are expressed in terms of atomic mass unit scale where *one atomic mass unit* (amu) corresponds to one-twelfth the mass of the carbon-12 isotope or 1.6604×10^{-24} g. In terms of this scale, the masses of the hydrogen atom and subatomic particles are,

Hydrogen atom 1.0078252 amu

Electron 5.4857×10^{-4} amu

Proton 1.0072766 amu

Neutron 1.0086654 amu

The total mass of an atom in terms of the individual masses of the building blocks of protons and neutrons and electrons is given by $ZM_H + (A-Z)M_n$ where M_H and M_n are respectively the masses of a hydrogen atom (1 proton + 1 electron) and a neutron. For example, the mass of a helium atom comprising 2 hydrogen atoms and 2 neutrons should be

Mass of two hydrogen atoms	2×1.0078252 amu
Mass of 2 neutrons	2×1.0086654 amu
Total	4.0329812 amu

(Note: The mass of electron is relatively negligible and hence ignored)

The exact mass of helium-4 however, as determined by mass spectrometry, is 4.002603 amu, indicating a mass deficit of 0.030378 amu. This deficit, though apparently small, assumes significance in terms of Einstein's Law of Conservation of Mass and Energy, which states that there exists an equivalence between the mass m and the energy E as given by the equation $E=mc^2$ where c is the velocity of light

(2.998 x 10¹⁰ cm). In terms of this relationship, the energy equivalence of 1 amu of mass works out to 931.5 MeV. The mass defect of 0.030378 amu in the formation helium-4 thus represents an energy equivalent of 28.2 MeV. In other words 28.2 MeV of energy has been released in the formation of one atom of helium. This energy, known as the *binding energy* (BE) of the atom, is supplied by the loss of the 0.030378 amu unit of matter, liberating 28.2 MeV of energy that holds the nucleus together. For splitting the helium atom into its component units, an equivalent amount of energy (i.e. 28.2 MeV) has to be supplied. Nuclear binding energy values are thus a million times greater than the energies involved in chemical reactions.

The actual mass of the atom of any element is consequently less than the sum of the masses of its constituent particles. For example, the actual masses of beryllium-9, Ca-40 and uranium-238 are less by 0.0624418 amu, 0.367223 amu and 1,9333 amu respectively with corresponding binding energy values of 58.16 MeV, 342 MeV and 1802MeV.

But the value that is more frequently considered is the *average binding energy per nucleon* (BE/A). This is obtained by dividing the total binding energy by the number of nucleons in the atom. The higher the average nuclear binding energy, the greater is the stability of the nucleus. Fig.2 shows the average BE/A values for various atoms as a function of their mass numbers (number of nucleons in the nucleus).

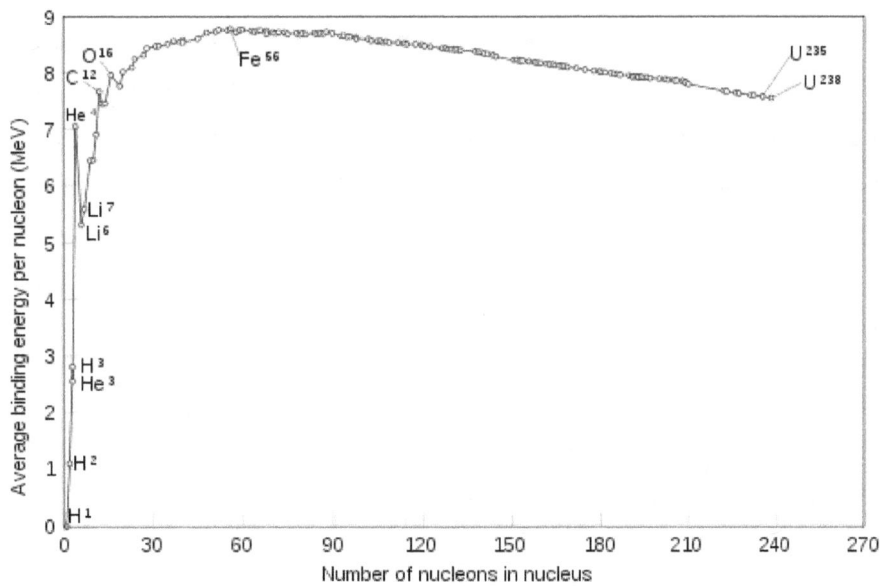

Fig. 2: Average BE/A vs. mass number

The BE/A values are fairly small for all nuclides except for a few of the lighter ones with their range lying between 7.6 MeV and 8.8 MeV. The dissociation of a heavy nucleus into lighter nuclei, called *fission*, takes the system from a low BE/A value to a higher value accompanied by the liberation of energy. For example, when uranium-235 with a BE/A value of 7.6 MeV is bombarded with thermal neutrons, the compound nucleus uranium-236 undergoes fission into two fragment elements, each with a BE/A value of 8.5.

As a consequence, the energy released per fission is 236 (8.5–7.6) = 212 MeV. If a mole (235 g) of uranium-235 is made to undergo complete fission, the energy released will be of the order of $212 \times 6.2 \times 10^{23}$ MeV or 4.4×10^{12} kcals or 2.2 MW of electric power/day. In contrast, the energy released when one mole of carbon (12 g) undergoes combustion forming one mole of carbon dioxide the energy released is a mere 94 kcals.

It can be seen from Fig.2 that when two nuclei of a lighter element (e.g. deuterium) can also combine to form a heavy nucleus with a lower average binding energy releasing energy. Such a reaction is called *nuclear fusion*. Energy liberation occurs in the fusion process as well. It is remarkable that nature has gifted man the isotopes of the heaviest (uranium) and the lightest (hydrogen) naturally occurring elements and they can be exploited as commercial energy sources through fission and fusion.

3.6 Nuclear Fission Chain Reaction

The most significant fact in a fission chain reaction is that more than one neutron is liberated in each fission event. These neutrons can cause further fissions if they do not escape from the fissionable material or are not captured by other materials. In such an event, the fission reaction becomes a self-sustaining chain reaction. Assuming that in a supercritical assembly two neutrons are liberated in a fission event, and each of these neutrons, in turn, produces two more neutrons, there will be 2^n neutrons at the end of n fission stages in a chain reaction (Fig.3). Let us assume that the average time between the emission of a neutron and its interaction with another uranium nucleus, causing fission, is of the order of a 10^{-8} sec (10 nanoseconds). In one second, neutron multiplication goes through 10^8 stages. This rapid multiplication of neutrons from a large number of fissions in such a short time leads to energy release with explosive violence. Leo Szilard was probably the first scientist to conceive how an atom bomb might work through a rapid chain reaction.x larger

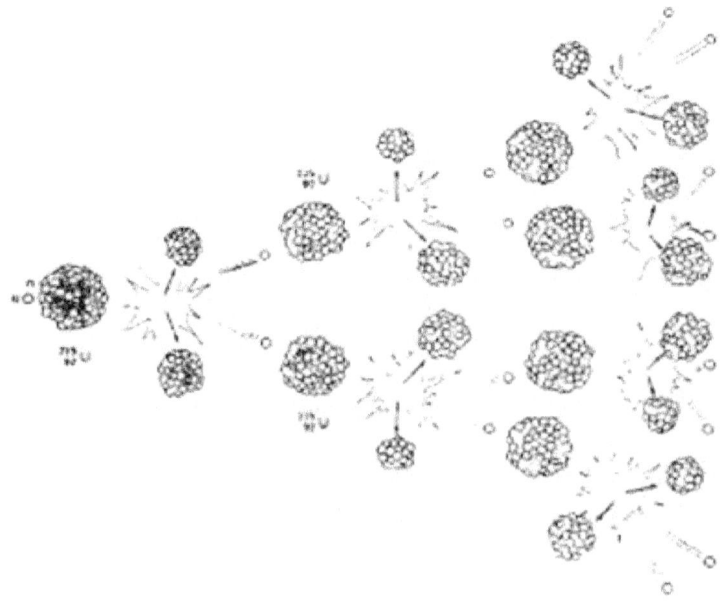

Fig. 3 Nuclear Fission Chain Reaction

3.7 Critical Assembly

A critical mass is the smallest amount of fissile material needed for a sustained nuclear chain reaction. For example, in a small-sized sample of ^{235}U, the probability of a large fraction of the fission neutrons escaping through its surface and thus, not becoming available for the propagation of the chain reaction is high. Such an assembly of fissionable material (uranium-235 or plutonium-239) is said to be a *subcritical assembly*. As the size of the sample increases, the fraction of the escaping neutrons decreases due to the lower increase in the surface area (proportional to πr^2) when compared with the volume (proportional to πr^3) and a size is reached where a chain reaction can be sustained through the availability of marginally more than one neutron after each fission. Such an assembly is said to be a *critical assembly*. A nuclear reactor is a critical assembly of nuclear material, which can initiate and sustain just a self-sustaining fission reaction providing energy. When the size of the sample increases further, it leads to the condition where well above one neutron is available after each fission. This results in a rapid increase in the number of fissions in successive generations of fissions. Such an assembly is said to be a *supercritical assembly*. An atom bomb is a super critical assembly of a pure fissionable material such as U^{235} or Pu^{239} in which a chain reaction is initiated for the rapid and uncontrolled multiplication of fission reactions releasing huge quantities of energy in a fraction of a second. A neutron reflector, a light element like hydrogen or beryllium, reduces the mass needed for criticality. A neutron reflector, a light element like hydrogen or beryllium, reduces the mass needed for criticality. The U^{235} isotope present to the extent of 0.72% in natural uranium is enriched to the required level while Pu^{239} is synthetically prepared by reaction between thermal neutrons and U^{238} in reactor according to the reaction,

$$U^{238} + n^1 \rightarrow U^{239} \rightarrow Np^{239} + \beta^- \rightarrow Pu^{239} + \beta^-$$

Plutonium-239 is an α-emitter with a half-life of 24,200 years. Yet another isotopic species that undergoes fission with thermal neutrons is U^{233}. It is prepared by thermal neutron reaction with thorium according to the reaction,

$$Th^{232} + n^1 \rightarrow Th^{233} \rightarrow Pa^{233} + \beta^- \rightarrow U^{233} + \beta^-$$

The numbers of neutrons produced in the thermal neutron fission of each of these fissionable isotopes are,

Uranium-235 2.40

Plutonium-239 2.80

Uranium-233 2.50

All these isotopes undergo fission with fast neutrons also, but the fission rate decreases as the energy of the incident neutrons increases.

3.7.1 Spontaneous Fission

Spontaneous fission is a naturally occurring nuclear decay process exhibited by a few heavy elements starting with uranium and beyond. The nuclei of these elements, which are naturally unstable, also undergo fission by themselves because their nuclei are so massive that their binding energy cannot hold them together indefinitely. The nuclei split with characteristic half-lives into two approximately equal parts, as in induced fission, releasing one or more neutrons. Consequently, a limited number of neutrons are always present in these elements as well as in a reactor core that has never been operated. These neutrons released in spontaneous fission initiate all nuclear chain reactions in nuclear weapons and nuclear reactors when the fissionable material is assembled to its critical size. The time required for each generation of neutrons is determined as 8×10^{-8} second. Thus, in a second a single neutron with a multiplication factor of two, after going through 80 generations cumulatively generates 1.24×10^{24} neutrons, creating conditions for an explosion. For this reason, great care should be exercised in handling the fissile material so as prevent it from attaining critical size. Such as in the processing and fabrication plants that handle fissionable materials including highly enriched uranium and plutonium, either in solid form or solution form.

At Los Alamos in 1944–45, Otto Frisch was stacking blocks of enriched uranium to conduct an experiment without a neutron reflective assembly. In the course of the experiment, he leaned over the blocks to give instruction to a nearby graduate student. The red light on the neutron counters started glowing continuously. Frisch realised that the white cloth of his lab coat and the water inside his chest were reflecting neutrons back into the uranium blocks creating conditions of criticality. Frisch immediately knocked several of the blocks onto the floor with his forearm. If the exposure had been a little longer the criticality condition would have led to his instantaneous death!

3.8 Use of Atom Bombs in World War II

The first atom bomb, Little Boy, dropped over Hiroshima was a gun device in which two pieces of highly enriched uranium, each less than a critical mass, were brought together very rapidly to form a single supercritical assembly. This is achieved with a tubular uranium device in which a highly enriched piece is held at the opposite end of the tube (Fig.4).

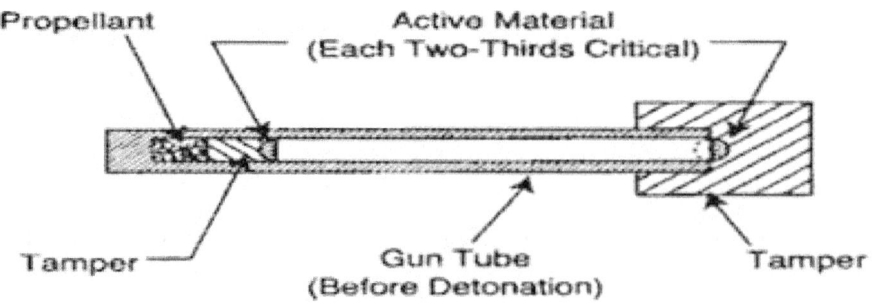

Fig. 4: Uranium Bomb (Little Boy) (Gun Assembly)

The second bomb, Fat Man, dropped over Nagasaki was an implosion type assembly of plutonium in which high explosives are arranged to form an imploding shock wave which compresses the fissile material to supercriticality (Fig.5).

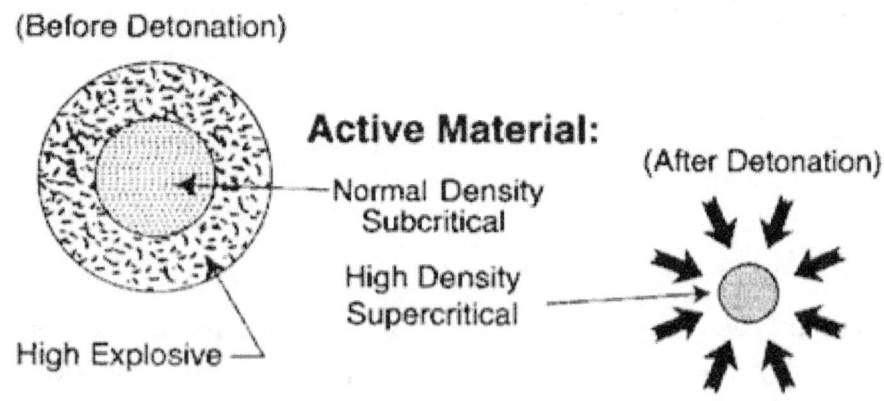

Fig. 5: Plutonium Bomb (Fat Man) (Implosion Assembly)

3.9 Fission Products

Nuclear fission products are the atomic fragments generated after the fission of U^{235} or Pu^{239} nuclei. As most of the fission fragments are neutron-rich, they decay by β^- and γ emission to stable nuclei. During the fission of U^{235}, about 1000 unstable and stable fission products are formed. The peak fission yield corresponds to isotopes with mass numbers 95–100 and 135–140. A considerable number of fission products such as Sr^{90}, Cs^{137}, Tc^{99} and I^{129} are medium to long-lived isotopes.

3.10 Nuclear Fuel Reprocessing

An important part of the nuclear energy programme is the processing of the spent nuclear fuel. This recovers unused uranium and plutonium for recycling and removes the fission products, which act as poisons making the fission process less effective. The composition of a spent fuel depends upon the initial quantities of uranium (natural or enriched) and the time the fuel has been in the reactor. For example, the used fuel from light water reactors using enriched uranium contains approximately 95.6% uranium, 2.9% stable fission products, 0.9% plutonium, 0.4% Sr, Cs and I fission products, 0.1% other long-lived fission products and 0.1 minor actinides. After separation, the high-level fission product wastes are concentrated, immobilised in highly inert matrices such as glass, and sealed in corrosion-resistant containers. After keeping them for a sufficient amount of time for the radioactivity to decay, these containers are buried in deep geologic repositories.

4

Uranium Prospecting and Mining

4.1 Uranium Rush

As work at the Manhattan Project progressed, there was a steep rise in demand for uranium. In August 1943, General Groves, who was in charge of the Manhattan Project, commissioned a team for the evaluation of the world's uranium resources. The team reported that there was just enough uranium in the US to see the project through with hardly any uranium left to pursue future goals of power generation or build an arsenal of nuclear weapons. Groves immediately initiated steps, with the help of the US Army, to get hold of the stocks of uranium lying in the Belgian Congo. After the conclusion of World War II, a joint development agency comprising the US, UK and Canada began exploratory activity for uranium in their countries as well as in other friendly countries.

The post-war decade turned out to be a period filled with a rush for uranium. *"Every American oil-man looking for 'black gold' in a foreign jungle is derelict in his duty to his country if he has not at least mastered the basic information on the geology of uranium,"* declared Gordon Dean, the Chairman of the US Atomic Energy Commission. Encouraged by such statements and the incentives offered by the US Government, many professional and amateur explorers and enthusiasts, raced around in search of the metal, carrying GM counters. Intense activity was witnessed in Canada, Australia, and other countries. These explorations unveiled substantial uranium deposits in Canada, Australia, France and French African colonies like Niger, Gabon, and Namibia. In South Africa, uranium was found in gold ores. The uranium ore, pitchblende, was found in central France.

The USSR did not have domestic stocks of uranium when it initiated its own atomic energy programme. The Red Army took over St. Joachimstahl, the birthplace of uranium mining and using forced labour, comprising convicts and prisoners of war and others, started uranium mining on a war footing. Mining operations were also extended to Erzgebirge Mountains in East Germany, where uranium deposits were discovered. Explorations undertaken simultaneously in the USSR yielded good results.

Uranium mineral exploration in India began in 1948 under the exclusive control of the Atomic Energy Commission, constituted by the Government of India. All the deposits found so far are a low grade.

4.2 Uranium Prospecting

Uranium, a metal almost as common as tin or zinc, mostly occurs in low concentration (a few parts per million) in soil, rocks and water. Some typical concentration levels are,

- Very High Grade Ore (Canada)-20% U 200,000 ppm U
- High Grade Ore – 2%U 20,000 ppm U
- Low-Grade Ore - 0.1% U 1,000 ppm U
- Very low-grade Ore – 0.01% 100 ppm U
- Granite 4.5 ppm U
- Sedimentary rock 2.0 ppm U
- Earth's continental crust (av) 2.8 ppm U
- Seawater 0.003 ppm U

Source: World Nuclear Association, 2010.

Ores containing 1,000 ppm (0.1%) or more uranium are considered as commercial grade. However, some countries process even lower grades (about 0.05%) to meet their needs (e.g. India).

4.2.1 Preliminary Survey

Prospecting for uranium is conducted very much like exploration for other minerals. Additionally the radioactivity of this element makes it amenable for its detection through the use of special instruments (e.g. GM Counters).

The first step involves the identification of the geological environments conducive for the occurrence of the element. For this, the prospector needs a knowledge of local geology and ore deposits.

It is now known that uranium occurs in a variety of igneous, metamorphic and sedimentary rocks, ranging in age from less than one million years to over three billion years. Geological mapping and analysis by conventional methods help in the identification of such zones. Satellite imagery and aerial photography (remote sensing) also provide useful background data. Remote sensing involves the analysis of the ground surface of the zone using sensors onboard airborne (aircraft) or spaceborne (satellites) platforms. Starting with LANDSAT launched in 1972 in the US, a number of satellites have been used to study the surface land features, geologic structures such as folds and faults and prepare rough geologic maps of large areas. This activity is not possible through traditional surveys. These pictures show the mineral concentration zones in distinct colours. India's first remote sensing satellite 'Resourcesat-1' was launched in 2003. This is followed by Resourcesat-2 in April, 2011.

4.2.2 Identification of Anomalies

The time gap between a preliminary survey and authentication of uranium mineral deposits in a broad area may be a decade or more. This is followed by narrowing down the target area through radiometric surveys. The GM counter was the first instrument, in the 1930s, available for this purpose. This portable instrument (Fig.1), however, had the limitation of not being able to discriminate between the radioactivity of uranium, thorium, and potassium. The γ-ray spectrometer developed in the 1940s has proved a more versatile and valuable tool (Fig.2) for the purpose.

For surveying limited areas, the best method consists of a slow walk over the ground with a radiation detector held at a constant distance from the ground. Some instruments have built-in audio devices to alert the explorer. If the radioactivity at any place reads four times or more than the background value, a sample is collected for a detailed laboratory study. A radioactive bed or vein thus, identified is then subjected to systematic measurements such as thickness and strike length. Basing on the data, a rough estimate of the uranium content of the deposit is arrived at.

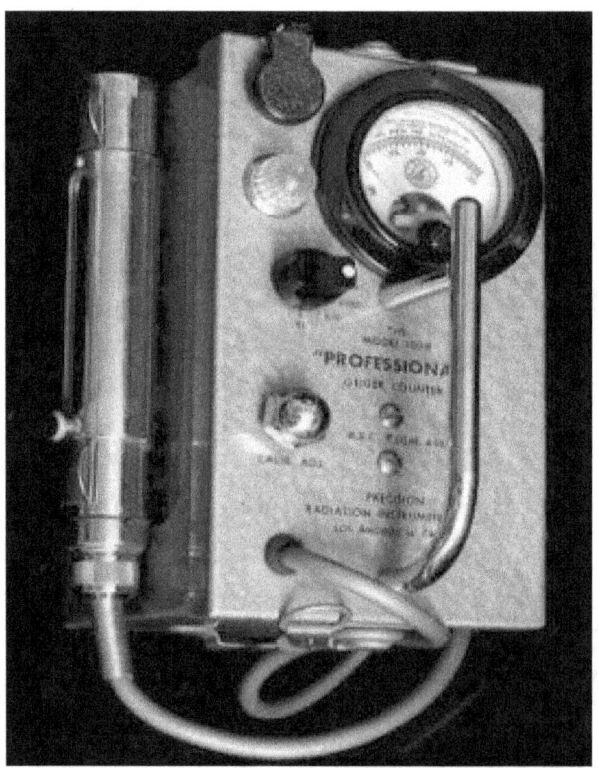

Fig. 1: GM Field Detector

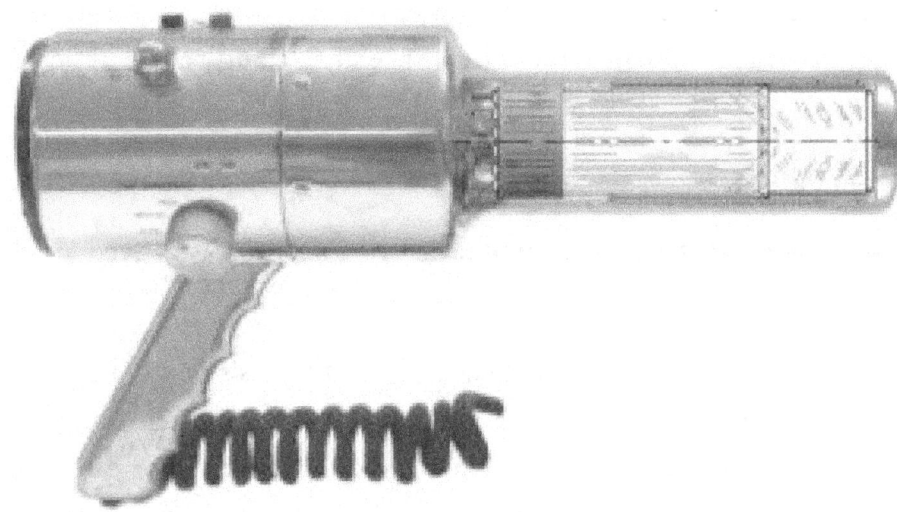

Fig. 2: Scintillation Detector

The radon measurement method is helpful for locating deep deposits of uranium ore, Radon gas is an alpha-emitting daughter product of uranium decay (through radium). This gas, which diffuses through the rock fissures and soil, can be detected using an alpha sensitive photographic track detector exposed to the soil gas for a known length of time. The density of the alpha particle tracks gives a measure of the radon gas in the soil and thence a measure of the uranium ore deposit.

4.2.3 Geochemical Prospecting

Since 1950, a number of geochemical methods have been developed for mineral explorations. These methods are based on the principle of identifying the metallic species released into the surrounding environment (e.g. aquatic systems) through weathering processes. In the present context, uranium is converted to a water-soluble form under oxidising conditions. The metallic species is carried slowly by groundwater into the surrounding soil. Depending on the acidity or alkalinity of the soil and the presence of reducing substances, uranium gets concentrated and precipitated in the soil. Analysis of soil, stream and lake waters, and sediments will reveal these anomalously high uranium contents (in parts per million range), which in some situations could help identify potential deposits in the area.

4.2.4 Evaluation of a Uranium Ore Deposit

The identification of a promising source is followed by an in-depth survey with respect to its size (in terms of available quantity of ore), shape and depth and most importantly its average uranium content. Such a survey is based on the mineralogical and chemical analysis of the samples obtained through drilling operations. Since drilling is an expensive operation, the choice of sites for drilling holes has to be made judiciously. The samples obtained from the drilling operations are then subjected to the extraction of uranium by conventional methods at different scales (e.g. laboratory, pilot plant) of operations.

These factors determine the cost of recovery of uranium as a concentrate (Yellow Cake) from the deposit. The presence of other recoverable minerals in the deposit will also add value to the deposit.

4.3 Mining of Uranium

An ore body goes through several stages of appraisal before a decision is arrived at for its mining. These include:

Economic viability: Operational and processing costs form the basis of any large-scale mining venture. These, however, assume secondary importance in situations where a country's nuclear energy programme faces problems with no access to uranium in the international market. For example, India's ambitious nuclear energy programme is beset with this problem. It has only limited domestic low-grade uranium resources. With no access to uranium in the international market due to sanctions (these are now relaxed), India had no alternative but to process its low-grade ore, despite high costs. However, the uranium thus, recovered could only partially meet the country's targets of power production.

Adverse impacts: In many countries, including India, uranium mining is the subject of heated debates with civil society objecting to this activity on the grounds that it poses an adverse environmental impact on the area.

4.3.1 Mining Techniques

Uranium mining marks the beginning of the *nuclear fuel chain*. This chain represents the progression to nuclear fuel fabrication (e.g. uranium) through a series of steps starting with the mining of uranium and ending with the disposal of nuclear waste (Fig. 3).

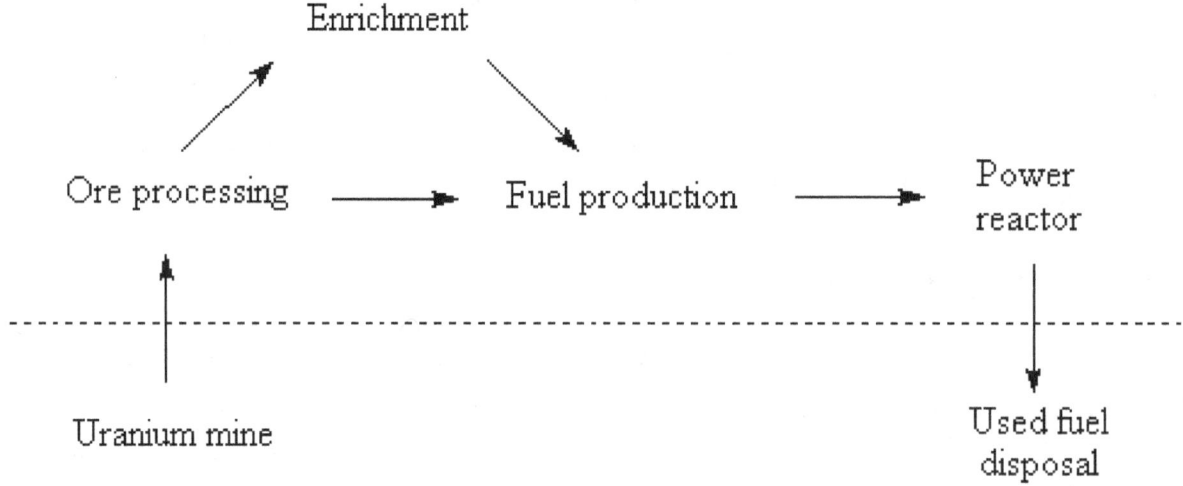

Fig. 3: Nuclear Fuel Chain

Since uranium mostly occurs in low concentrations (~ 0.1%) large quantities of the rock have to be mined for extracting a tonne of the product. Added to this, the radioactivity of uranium creates problems in mining and handling of tailings from processing plants. These will be considered in a later chapter.

Uranium is also recovered as a byproduct in some parts of the world. An important example is the Olympic Dam in Australia where copper is the main product. In South Africa, uranium is recovered as a minor constituent from gold bearing ores. In these cases, the concentration of uranium may be as low as a tenth of the quantity occurring in primary ore bodies. In the US, uranium was obtained for a long time (in the 1940s and 1950s) as a byproduct from vanadium-bearing carnotite ores in Colorado, US.

4.3.1.1 Open-Cast Mining

Ore bodies close to the surface are generally mined by the 'open-cast' or 'open-pit' technique. The topsoil or rock that covers the deposit is removed by blasting followed by the use of draglines or excavators. A lot of waste rock may also have to be cleared while excavating the ore. The waste rock is usually stacked near the mine and is used for filling the pit back in after mining. The ore is graded for its uranium content on the basis of its radioactivity.

Open-pit mining is more cost-effective compared to underground mining. Almost 30% of the world's uranium comes from open-pit mines. In open-pit mining, quick dispersal of the hazardous radon gas (the product of uranium decay) takes place causing less risk to mining personnel. Water is extensively sprayed on the ore during mining to control the radioactive dust.

Some of the important open-pit mines in the world include Rangers in Australia, Rossing in Namibia and most of the mines in Northern Saskatchewan in Canada.

Source: UCIL

Fig. 4: An Open-Cast Uranium Mine in India

4.3.1.2 Underground Mining

Underground mining is used to for gain access to deep ore deposits. Access to the uranium ore formation is achieved through digging tunnels and shafts. In underground mining, less waste rock is removed. As a result, there is a less adverse environmental impact. But the radon gas which is confined in the tunnels presents a hazard to miners. In this case, protection is provided through efficient ventilation.

Many of the major uranium mines of the world are in the underground category (Figs. 5 and 6). The Olympic Dam in Australia (the biggest uranium-bearing mine in the world), McArthur River, Rabbit Lake and Cigar Lake in Saskatchewan in Canada, Akouta in Niger in Africa are some examples. The low-grade mines in Jaduguda, Bhatin, Narwapahar, Turamdih, Baghjata, Banduhurang and Mohuldih in Jharkhand, India are also underground. A recent addition to the list is Thummalapalli in Andhra Pradesh, India.

Source: UCIL

Fig. 5 Underground Uranium Mine. Narwapahar

Source: UCIL

Fig. 6: Entrance to an underground Uranium Mine

4.3.1.3 In Situ Leaching

In situ Leaching (ISL) or solution mining is used for uranium extraction from the ground without mining the ore. This is achieved by directly injecting chemical leaching reagents into the ore body. Since this technique works with permeable uranium-bearing strata only, sandstone deposits in aquifers are suitable for this type of mining. Geologically these deposits are confined above and below by impermeable strata of clay (called aquicludes), which act as barriers, preventing the leachates from diffusing beyond the leaching zone and contaminating the groundwater.

The technique involves the injection of weakly acidic or alkaline leaching solutions into different parts of the ore body through injection wells. Sodium carbonate is used as a leaching agent if acid-consuming minerals like carbonates of calcium or magnesium are present in an appreciable concentration in the ore. Otherwise, dilute acid is used as the leaching agent. An oxidising agent like oxygen or hydrogen peroxide is also added. The leaching solution slowly migrates through the aquifer, leaching the uranium-bearing sand. The leachate is collected in strategically placed recovery wells (Figure 7). The uranium-bearing liquid is then pumped to a chemical plant for extraction of uranium. Normally, the ion-exchange method is employed for this purpose.

After extraction of uranium, the uranium-free solution is reconditioned through the addition of reagents and partly fed back into the injection wells. In a properly operated system, the solution pumped out will be more than the volume of solution pumped in through the injection wells. This will keep the leachate flowing towards the extraction wells rather than drifting across the leaching boundaries.

The ISL technique is cost-effective since there is no need for excavation and processing of huge quantities of ore. Another major advantage is the relatively low environmental impact as compared to the open and underground operations. The absence of bulky tailings for disposal is another plus point.

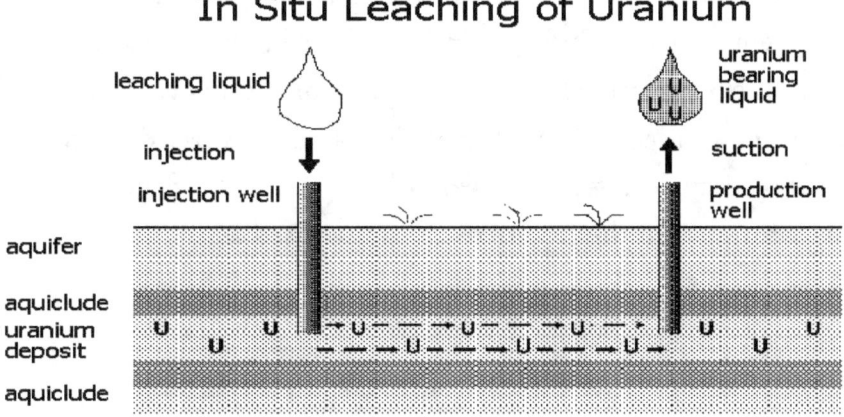

Fig. 7: In Situ Uranium Leaching

Notwithstanding all these advantages, several objections are raised against the ISL process. One is the risk of many toxic metals like thorium, radium, lead, arsenic, and cadmium leached along with uranium reaching the ground surface. These have to be dealt with as part of the waste stream.

The second is the fear that the uranium-bearing solutions might escape from the ore zone through the vent or fault lines, entering the aquifer system and polluting even the groundwater away from the mine site.

The third is the problem of disposal of large volumes of wastewater generated in the processing plant in an environmentally benign manner.

These objections are met by the stipulation that the mine operator, after completion of the mining operations, should restore the ground quality to its original condition as specified in terms of chemical analysis. When this is effectively implemented the footprints of the ISL operation will have been almost completely erased. The barren liquid after uranium recovery is also treated through reverse osmosis to remove the dissolved salts. Part of this clarified water is allowed to flow freely over the ground irrigating fields. Water samples drawn from the wells located in the vicinity of the ISL mine are regularly monitored for any leakage of fluids from the mining zone. Laxity in the enforcement of these regulations is often the genesis for protests by civil society against mining operations.

About 20% of the world's uranium comes from ISL operations. Beverley in Australia was the first mine to use ISL in 2000. Other mines like Honey Moon, Australia, are also scheduled to use ISL. Smith Ranch-Highland mine in Wyoming and the Crow Butte mine in Nebraska (both in the US) are smaller operating units. Kazakhstan extracts most of its uranium using this method.

4.4 World Uranium Resources

The World Nuclear Association (WNA) and OECD (Organization for Economic Development and Cooperation) Nuclear Energy Agency (NEA) jointly with IAEA regularly provide data on resources of uranium in each country. The publication from NEA is popularly known as "Red Book." The methodology of reporting the data is somewhat complicated. Simply stated, it is based on the confidence level of the estimates communicated by each country and the market-based cost of producing Yellow Cake from the resource. Based on this information, the resources are categorised as:

- Reasonably Assured Resources (RAR)
- Estimated Assured Resources – Category-I (EAR-I)
- Estimated Additional Resources-Category-II (EAR-II)
- Speculative Resources (SR)

Reasonably assured sources (RAR) represent established resources. The rest of the known resources categorised as EAR-1. EAR-II and SR include resources known with less certainty. The IAEA encourages all member countries to report their uranium resources using the above classification or any other easily comparable one. In each category, particularly in RAR and EAR-I, the resources are subdivided into three price bands: <US$ 80 per kg U, US$ 80–120/per kg U and US$ 130 per kg U. If we accept to pay a price higher than US$ 80–130/kg then more resources can be found. These price bands have varied with time over the past few decades.

The NEA/IAEA scheme considers only the lowest cost RAR (recoverable at US$ 80 per kg U or less) as reserves. The idea is to identify reserves that are of economic interest at the time of assessment. The known recoverable resources of different countries as of 2009 and reported by WNA are shown in Table-1. The total world resources at <130 US$/kg U price level are estimated at about 5.5 million tonnes of uranium. Nearly 70% of world resources are located in six countries – Australia, Kazakhstan, Russia, South Africa, Canada, and the US with 20–25 other countries sharing the rest (Table 1).

4.4.1 Australia

Australia, which has been engaged in active exploration from the 1960s, is the leading supplier of uranium. The Olympic Dam underground mine, discovered in 1975, commenced production in 1988. Considered as the largest uranium resource in the world, it has a massive deposit lying 350m below ground. Though it is low grade, the uranium recovery is economical due to the byproducts copper (major), silver and gold (minor). Uranium accounts for 20% of the revenue, copper 75% and silver and gold 5%.

The Ranger mine in the Alligator River Region is an open-cast mine. Discovered in 1969, it commenced operation in 1980 and reached full production in 1981. The original ore body was completely mined out by 1995. A second ore body began mining in 1997. Deposits have also been identified at Jabulika, Koongarra, and Nabarlek. The Beverley mine, the first commercial venture, is a relatively young sandstone deposit lying at a depth of 100–130m.

The total production from the three mines in 2007–08 was about 8,560 tonnes of uranium. In addition, Australia has identified deposits at Angela, Kintyre, Yeeleerrie and Mulga Rock.

Table1: Known Recoverable Resources of Uranium (2011)

Country	Uranium (tonnes)	Percentage
Australia	1,661,000	31
Kazakhstan	629,000	12
Russia	487,000	9
Canada	468,700	9
Niger	421,000	8
South Africa	279,100	5

Continue...

Brazil	276,700	5
Namibia	261,000	5
US	207,400	4
China	166,100	3
Ukraine	119,600	2
Uzbekistan	96,200	2
Mongolia	55,100	2
Jordan	33,800	1
Other	164,000	3
World Total	**5,327,200**	**100**

Source: World Nuclear Association, 2012.

4.4.2 Canada

Canada's uranium mining operations began in the 1930s with the extraction of radium from Port Radium deposits. In the wake of the US war-time Manhattan Project, the mining activity intensified and by 1956, thousands of radioactive anomalies were identified. Within three years, 23 mines, big and small, were in operation. By 1960, Canada was earning more from uranium ore than from any other ore. The production peaked at around 12,000 tonnes with the main activity centered in Elliot Lake Area, Ontario. By the 1970s, Saskatchewan in Athabasca basin came into prominence with deposits at Rabbit Lake, Cluff Lake, Key Lake, McLean Lake, and Cigar Lake. In 1988, the massive McArthur River deposit, which was high-grade (over 20%), was located at a depth of 600m. Cigar Lake (400m deep) is another high-grade deposit. As these high-grade ores posed serious radiation hazards for miners, special strategies for mining had to be designed. For example, at the Cigar Lake deposit, ground-freezing and remotely-operated high-pressure water jets are used for excavation. The high-grade ore in a slurry form was trucked to the McLean Lake mill to produce 7,000 tonnes of uranium per year. Canada invested $539 million between 1980–97 on uranium exploration.

4.4.3 US

Regular production of uranium-bearing ores in the US began in 1898 when carnotite (a uranium-bearing mineral) bearing sandstones in Colorado and Utah were mined for their vanadium value. With the discovery of radium, exploitation became more attractive. Uranium became a byproduct, but with a limited market. By 1913, the Colorado Plateau was catering to about half the world's radium demand. Production declined sharply after 1923 when high-grade uranium ores in the Belgian Congo became a more viable source of radium. Added to this vanadium from Peru captured the market resulting in the closure of US deposits.

The uranium byproduct from the above-mentioned mines as well as the indigenous carnotite proved very handy when the Manhattan Project was started. The mining companies were, however, unaware that the US government was buying carnotite from them for the extraction of uranium. The late 1940s and early 1950s saw a boom in uranium mining in the Western US, with some individual prospectors making huge fortunes. Other known uranium deposits in the US are distributed in the states of Wyoming, New Mexico, Arizona, Colorado, Utah, and Texas. With the opening of many deposits in Canada and Australia and a decline in uranium prices in 1992, there was a sharp slump in US uranium production. At present, only a limited amount of uranium is produced in Wyoming and Nebraska through ISL using carbonate leaching. The US currently meets its nuclear fuel requirements mostly through imports of fuel material (plutonium and enriched uranium) from dismantled nuclear weapons.

4.4.4 South Africa

South Africa has the world's fourth-largest uranium reserves with an estimated recoverable quantity of about 350,000 tonnes of uranium. A large part of this is the byproduct from gold tailings at a concentration of about 0.01%. In the early years, South Africa was a major supplier of uranium, particularly to the UK. In the gold belt of Buffelsfontain basin as many as 26 mines were in operation, between 1945–65, for uranium extraction. When production was at its peak 6,100 tonnes of uranium were produced annually. By 1983, the number of mines fell to 14. By 2006, only four mines – Vaal River, Palabora, Hartebeesfontain and Western Areas – were in operation. Currently, only the Vaal River plant is producing 1,270 tonnes uranium a year. An attractive price for uranium could increase the country's production once again.

4.4.5 Kazakhstan and Russia

Kazakhstan and Russia jointly control a quarter of the global recoverable uranium resources. Speaking at the World Nuclear Association's Annual Symposium in 2004, Mukhtar Dzakishev, President of KazAtomProm, declared, "Kazakhstan can and should be one of the key sources for supplying world nuclear energy with natural uranium."

The state-owned National Nuclear Company KazAtomProm is the sole authority for uranium mining, reprocessing, export, and import operations. The total resources and reserves of uranium in the country are put at 1.5 million tonnes out of which a major part, about 75%, is recoverable through the relatively low-cost ISL technology.

At present, Kazakhstan is the only country where ISL technology dominates. In the Shu-Saryusa province, uranium production is carried out in the Uranas, Eastern Mynkuduck, Kanzugan, and Southern Monikun deposits. Mining is also ongoing in Inkai and Moinkum in the northern and southern Karamurun deposits. A joint venture with Russia is slated to start production at the Zarechnoye deposits.

With an expected surge in demand from China, India, and Japan for uranium, Kazakhstan plans to raise its mining capacity to about 12,000 tonnes from the 2008 level of 8,600 tonnes. Kazakhstan has entered or is entering into joint ventures for uranium production with South Korea, China, and Japan, mostly using ISL technology. The Russian State enterprise, Atomrednet – Zolota (ARMZ) plans to raise its output to 4,300 tonnes in 2009 from the 2008 level of 3,850 tonnes. These two countries are also aiming at higher targets of 20,000 tonnes by 2024–2025.

4.4.6 India

The Government of India established, as early as 1948, the "Rare Minerals Survey Unit" as part of the Department of Atomic Energy (DAE), for undertaking a survey of mineral deposits of importance in the country's atomic energy programme. This unit, which started functioning in New Delhi was later renamed "Rare Materials Division, RMD" and "Atomic Minerals Division, AMD" (1958). Its headquarters was shifted to Hyderabad in 1974. On July 23, 1998, this organisation got its present name "Atomic Minerals Directorate for Exploration and Research, AMD." The Directorate's principal mandate is to carry out geological exploration for mineral deposits of uranium, thorium, beryllium, zirconium, niobium, tantalum, lithium and other metals required for the nuclear power programme.

4.4.6.1 Atomic Minerals Directorate for Exploration and Research (AMD), India

In addition to the centralised laboratories and specialised groups at its Hyderabad headquarters, the AMD has established regional centres in New Delhi (Northern Region), Jaipur (Western Region), Shillong (North-Eastern Region), Jamshedpur (Eastern Region), Nagpur (Central Region) and Bangalore (Southern Region). Two sectional offices are also functioning in Thiruvananthapuram and Visakhapatnam, mainly for beach sand and off-shore investigations for minerals like ilmenite and rutile (for titanium), zircon (for zirconium), and monazite (for thorium and rare earths).

Supported by adequate technical facilities, the AMD has been carrying out air-borne and ground radiometric, magnetic, geophysical and geochemical surveys in geologically and structurally favourable areas in the country. Exploratory drilling and sampling for assessment of resources are also undertaken. Over the past six decades, several uranium deposits have been identified. It met with success in exploring the following types:

Vein Type: Deposits of this type are located in Singhbhum District (Jharkhand), Aravalis (Rajasthan), and Bodal-Jajawal (Madhya Pradesh). Of these, the Singhbhum Belt was extensively explored in 1950s. Currently, mining of ore and production of Yellow Cake are confined to this area. The mines presently in operation include: Jaduguda (1967), Bhatin (1986), Narwapahar (1998), Turamdih (2002), Bhagjata (2004) and Bandhuhurang (2006). Other potential areas are Mohildih, Nandup, Rajgaon and Giridih.

Sandstone Type: Deposits of this type are confirmed only in the North-Eastern parts of India (Meghalaya). Domisiat is a proven deposit. Another deposit at Wakhyn is under investigation.

Quartz Pebble Conglomerates: Exploration of this form of deposits, which yielded very rich resources like Elliot Lake in Canada, did not prove fruitful. Currently the Walkunji (Karnataka) and Dhanjori basins (Bihar) are under investigation for this type of formations.

Unconformity Related Type: This type is considered most promising. The lake deposits of Canada and Alligator River in Australia are good examples. In the 1990s, the AMD found uranium mineralisation in Andhra Pradesh along the unconformity between the Srisailam Formation of Cuddapah super group and the basement granites. Lambapur, Peddagattu, Thummalapalle deposits are in the YSR Kadapa District (earlier called Cuddapah District) in Andhra Pradesh. A mine is under development near Thummalapalle. The uranium content of these deposits is lower than 0.05%.

4.4.6.2 India's Uranium Resources

The Government of India announced in the Rajya Sabha on December 2, 2010, that as of October 31, 2010, India has 1,49,654 tonnes of uranium resources. The uranium deposits unveiled so far are low-grade (<0.05%) with no worthwhile byproduct recovery. The in-situ uranium content of the deposits is also not very high, usually around 10,000 tonnes or less of uranium. While most countries provide information on the cost of production no official information from DAE or AMD is available on the subject. A reasonable guess is that practically all known deposits fall in the category >130 US$/kg U.

The Domiasiat 8,000 tonne deposit in Meghalaya is reported to be of a higher grade with about 0.08% U. It is also possible to exploit this deposit via the open-cast method.

According to a news report in August 2012, India has discovered large deposits in Rohil in Rajasthan's Sikar District. At 5,185 tonnes, these deposits are considered the fourth largest in the country.

4.4.6.3 Uranium Mining in India

Mining of uranium is carried out exclusively by the Uranium Corporation of India (UCIL), a public sector enterprise under the DAE. The Jaduguda underground mine in Jharkhand was the first to start operation in 1967 with an initial capacity of 1,000 tonnes ore per day. In 1986 a small mine, 5km away, started supply of 100 – 150 tonnes ore per day to the uranium mill at Jaduguda. As the Jaduguda mine reached about 900m depth, its output started falling and the ore from Narwapahar mine was fed to the mill. At present, the total ore processed in this complex is about 2,010 tonnes per day, with an average of 0.037% U content.

To augment the domestic uranium output, the UCIL developed a second mining complex around Turamdih, about 24km from Jaduguda. The underground mines in Turamdih and Mohuldih and the open-cast mine in Banduhurang are expected to supply about 3,000 tonnes of ore of about 0.04% U, to a new processing plant at Turamdih daily.

A new mining project at Thummalapalle, in the YSR Kadapa District, Andhra Pradesh is expected to give a daily yield of about 2,000 tonnes of ore with 0.04% U content. The Thummalapalle mine has the potential to be the largest uranium mine in the world with a more than one lakh tonne deposit. If proven true the deposits will meet the fuel requirements of 10 indigenous reactors each of 700 MWe, coming up in the next 40 years, partly reducing the country's dependence on imported uranium. Production commenced in the mining and processing plant with a capacity of 3,000 tonnes per day in April 2012.

Though the Meghalaya deposits are of a somewhat higher grade (nearly 0.08% U) and can be mined via open-cast, no mining activity has been initiated so far due to strong opposition from the local population as well as the state government. There have been objections against exploiting the Peddagattu – Lambapur deposits in the Nalgonda District, Andhra Pradesh as well.

5

Nuclear Fuel Production

5.1 Yellow Cake

In the wake of the successful demonstration of a nuclear explosion, around a hundred uranium-bearing uranium ores were identified. The uranium ores are not mined for direct trade like other metal ores such as iron, tungsten, etc since most of the ores contain uranium ranging from 0.5 to 2.0 kg only per tonne of the ore. High-grade ores in Canada (with 10 to 20% uranium) are exceptions, but they are not commercially available for strategic reasons. Chemical processes were consequently, developed for the recovery of the element from some of these ores in the form of Yellow Cake, which is commercially available. Yellow Cake, which contains 70–80% uranium, is subjected to further refining to produce nuclear fuel. The processing of uranium ores for the production of nuclear fuel will now be described.

5.1.1 Crushing and Grinding

The ore, received from the mine in the form of lumps of about 20cm size, is crushed to less than 2.5cm size and then wet-ground into a fine slurry in a ball or pebble mills, making it amenable for extraction of uranium to over 90% through leaching operations.

5.1.2 Leaching

Leaching (also called lixiviation) involves the extraction of the desired component (uranium) from a solid matrix (the ore) by treatment with a suitable chemical solution. The leaching conditions depend on the composition of the ore and the mineral form in which uranium is present. Sulphuric acid or sodium carbonate is used as a leaching agent.

5.1.2.1 Sulphuric Acid Leaching

Sulphuric acid leaching is the most widely used process for the selective dissolution of uranium. At low concentrations, this acid dissolves uranium in the common uranium minerals like pitchblende (U_3O_8), uraninite (UO_2) and carnotite (potassium uranyl vanadate), leaving many of the accompanying elements present in mineral form more or less unaffected. The leaching conditions such as free acid concentration, temperature, and duration are adjusted to achieve recoveries of 90% or more of uranium. An oxidising agent is invariably added to convert the uranium to its hexavalent form to facilitate its dissolution. Manganese dioxide, in the form of the naturally-occurring pyrolusite, is often used for this purpose. Other oxidants that find use are sodium chlorate and hydrogen peroxide.

In exceptional cases of highly-closed ores such as brannerite from Elliot Lake, Canada, high temperature (75°C), high free acid concentration (50g per litre of free sulphuric acid) and long leaching time (>50 hr) are resorted to. The chemical reactions involved in the leaching step are:

$$UO_3 + H_2SO_4 \rightarrow UO_2(SO_4) + H_2O$$

$$UO_2 + [O] + H_2SO_4 \rightarrow UO_2(SO_4) + H_2O$$
(Oxidant)

$$U_3O_8 + [O] + 3H_2SO_4 \rightarrow 3UO_2(SO_4) + 3H_2O$$
(Oxidant)

Note: In high sulphuric acid concentrations uranium is present as the

Anionic complex $UO_2(SO_4)_2^{2-}$ or $UO_2(SO_4)_3^{4-}$.

The leaching operations are carried out in simple mechanically agitated tanks or pachuckas. The pachukas are tall tanks (Fig.1) in which air is introduced under slight pressure through a central tube lifting the slurry up and down to facilitate efficient mixing.

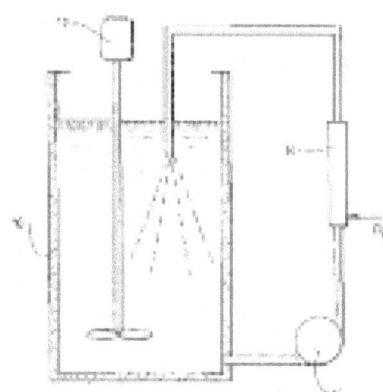

Fig. 1: Pachuka

Usually three or four mixing vessels are connected in a series. The slurry flows from one to the other before finally exiting to the solid-liquid separators.

5.1.2.2 Carbonate Leaching

Some ores containing carbonates of calcium and magnesium consume prohibitively large amounts of acid. In such cases, leaching is done using a mixture of sodium carbonate and bicarbonate.

$$UO_3 + 3Na_2CO_3 + H_2O \rightarrow Na_4UO_2(CO_3)_3 + 2NaOH$$

$$UO_2 + [O] + 3Na_2CO_3 + H_2O \rightarrow Na_4UO_2(CO_3)_3 + 2NaOH$$

Carbonate leaching has the additional advantage of being very selective for uranium. Other elements commonly present in the mineral are not attacked by carbonate. A high temperature is, however, required to achieve leaching at a reasonable rate. Air or oxygen under pressure acts as an oxidising agent. In some cases, leaching is done at higher pressures in autoclaves and higher temperatures (120° – 130°C). Currently, very few mills are operating with carbonate leaching.

5.1.3 Liquid-Solid Separation

After leaching, the uranium-bearing solution is separated from the solid residue. In most ores, the residue (called tailings) constitutes about 90% of the weight of original ore charge. The tailings are washed to remove most of the trapped uranium. Many devices are employed for this purpose. These include thickeners, cyclones, rake classifiers, drum filters, disc filters and horizontal belt filters used individually or in different combinations depending on the material handled as well as the economy. Counter-current devices are often used for solid-liquid separation as they are robust, reliable and easy to install and operate. However, such units are of large footprint needing provision of space and protection from heavy rainfall. Expensive filters are preferred for better recovery of uranium from high-grade ores. More recently horizontal belt filters, which provide very efficient washing, are coming into use. Minimising the loss of dissolved uranium at this stage is vital for minimising groundwater contamination as well as achieving maximum recovery of valuable uranium.

After passing through the filters, the uranium-bearing solution, called a *pregnant solution*, carries some suspended fine particles. As a very clear solution is required for the next stage of operation, it is subjected to clarification to the required level in mills employing continuous up-flow sand filters.

5.1.4 Concentration and Purification

At this stage, the uranium is present in the solution in low concentration, 0.5 to 2.5 grams per litre. Usually, sulphuric acid leachates contain a variety of impurities (mostly iron, aluminium, and silica) at low concentrations, depending upon the nature of the ore and leaching conditions. Carbonate leaching, on the other hand, contains less of impurities. Vanadium is a common impurity in the case of carnotite. Earlier, improved purity has been achieved by chemical precipitation methods. To meet the rising demand for uranium, these have been replaced with more efficient and cost-effective ion-exchange and solvent extraction methods.

5.1.4.1 Ion-Exchange

The ion-exchange process involves the interchange between the ions in an insoluble solid matrix and the desired ions in the electrolyte solution without a significant change in the nature of the solid matrix.

Zeolites (naturally occurring as hydrated sodium aluminosilicate minerals) have been in use as inorganic ion-exchangers for a long time for reducing the hardness of water. This involves the uptake of Mg^{2+} or Ca^{2+} in water by the zeolite through exchange with Na^+ ion in the solid matrix. Organic cation-exchange resins were introduced by Adams and Holmes in 1935. These are synthesised from styrene and divinyl benzene which form a polymeric network. This is then reacted with chlorosulphonic acid to introduce the sulphonic acid groups containing the exchangeable H^+ ion (Polymer-$SO_3^{-3}H^+$). When these resins were tried for extracting uranium in its cationic form (UO_2^{++}) from sulphuric acid leach solutions, it was found that adsorption was not only poor but was not specific, with several cationic impurities like iron (Fe^{3+}) also taken up by the resin.

In 1949, The Battelle Memorial Research Institute (USA) observed a preferential uptake of the anionic uranium species by strongly basic anion-exchange resins containing fixed quaternary ammonium groups {polymer-$N(CH_3)_3^+Cl^-$} and readily exchanging chloride ions. Using these resins, developed by Dow Chemical Company under a contract from US Atomic Energy Commission, the recovery of uranium in its anionic forms ($UO_2(SO_4)_2^{2-}$ or $UO_2(SO_4)_3^{4-}$ and $UO_2(CO_3)_2^{2-}$ or $UO_2(CO_3)_3^{4-}$) proved most promising. This method was first adopted in South Africa in 1951 for the large-scale production of uranium from gold tailings. By 1959, there were 17 plants in South Africa alone using this process for uranium production. By 1960, 24 mills in the US. Canada and Australia also switched over to this process. The technology was so efficient that uranium production reached a peak level of about 37,000 tonnes in 1959. The process also offered the advantages of separating uranium from many of the impurities present in the acid leach solutions and yielding uranium solutions of higher concentration amenable for separation by precipitation. Initially, the following three commercial resin products were used extensively:

Amberlite IRA 400, Rohm & Haas Co., US

Deacidite FF, Permutit Co., UK

Dowex-1, Dow Chemical Co., US

Over the course of time, resins were developed to take care of deleterious effects from impurities like iron and silica, present in fairly large concentration in some leach solutions. Currently, most ion-exchange operations use gel-type resins. Examples are,

Amberlite IRA-400, Rohm& Haas Co., US

Dowex 21-K, Dow chemical Co., US

Duolite 101-D, Duolite International, SA, France

Permutit SK, Permutit AG, Germany

Ionac 641, Sybron Corporation, US

The acid leach solutions normally contain 0.5–1.5g uranium, 20–30g sulphate, and 1 to 5g iron per liter of the solution at a pH of 1.5–2.0. Uranium is taken up from these solutions by the anion-exchange resins predominantly as the anionic complexes, $UO_2(SO_4)_2^{2-}$ and $UO_2(SO_4)_3^{4-}$. From alkaline leach solutions uranium is taken up as the anionic species, $UO_2(CO_3)_2^{2-}$ and $UO_2(CO_3)_3^{4-}$.

The ion-exchange process is generally carried out using fixed-bed column systems consisting of resin beads of -20 to +30 mesh size (about 0.45 mm.). The exchange capacity of the resin is around 30 g of U_3O_8 per litre of the resin. The column consists of,

A support at the bottom for the resin column,

Distributors at the top and the bottom for uniform

distribution of the feed solution through the resin bed,

Space above the bed to allow for the fluidisation of

The resin bed during back-washing, and

Piping, valves, and instrumentation to regulate flow of various solutions and also to control cyclic operation of process steps. Same size

A simple ion-exchange column for laboratory operation is shown in Fig.2

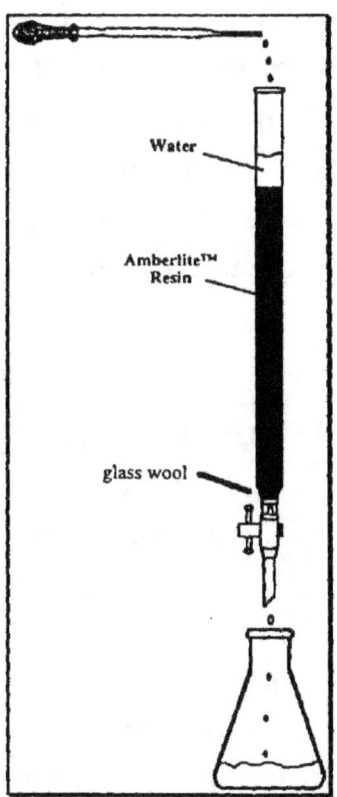

Fig. 2: Ion-exchange column

The cycle of operations in a typical ion-exchange process consists of the following steps:

Loading: The pregnant solution is passed through the column till the resin is loaded with uranium to the maximum possible extent.

Rinsing and back-washing: The excess solution remaining in the column is displaced and the bed is washed by up-flow of acidified water to expand and settle the resin. This step, called 'back-wash,' removes the suspended solids collected on the bed during the loading stage.

Elution: Displacement of adsorbed uranium from the loaded resin, called 'elution,' is carried out with 1 molar solution of sodium chloride or nitrate. This step also prepares the resin for the next uranium loading cycle. It is a common practice to collect the eluate in two nearly equal fractions. The second fraction, which is relatively dilute with respect to uranium, is used as the eluate in the first stage of the next elution cycle. This helps in a more effective use of eluant chemicals and also providing a solution of higher concentration to the precipitation stage. In some plants sulphuric acid is used for elution.

Washing: The excess eluant is displaced (removed) from the column by washing with water. This makes the column ready for next separation cycle.

A wide variety of both batch and continuous ion-exchange systems have been developed for achieving efficient anion-exchange separation of uranium.

There were some drawbacks in the use of ion-exchange process for the extraction of uranium from the leach liquors from different types of ores. When impurities like silica, titanium etc., present in the leach liquors get precipitated in the resin beads, the operation becomes sluggish. Added to this, the anionic species of some metals like molybdenum and cobalt (when present as cyanide complexes) get strongly adsorbed on the resin. As their removal is difficult, the resin capacity decreases gradually. The process was also time-consuming. Faced with these problems, the industry began looking for alternative technologies.

5.1.4.2 Solvent Extraction

In the early days of the Manhattan Project, extraction of uranium from nitrate solutions with diethyl ether was extensively used for large-scale uranium purification. Investigations showed that Tri-butyl phosphate (TBP), a neutral alkyl phosphate, is a more attractive solvent for this purpose. Though TBP as an extraction agent became popular in the nuclear industry, the requirement of highly-concentrated nitrate medium needed for extractions ruled out its use for sulphate or carbonate leach solutions. Studies in laboratories like Oak Ridge National Laboratory and other institutions, led to the development of more suitable reagents capable of extracting uranium from the leach solutions. Using these, it has also become possible to develop continuous multistage counter-current extraction systems with greater efficiency in terms of product purity and extraction efficiency.

The solvent extraction process in its simplest form consists of three steps:

Extraction: The aqueous uranium solution is intimately mixed with a suitable organic extractant when uranium passes with a high degree of selectivity into the organic phase.

Scrubbing: The loaded solvent is intimately contacted with a suitable aqueous solution to remove any minor impurities that are co-extracted, retaining practically all the uranium in the organic phase.

Stripping: The scrubbed solvent phase is then mixed with a suitable aqueous solution when the uranium is back-extracted into the aqueous phase for further processing. By a proper selection of the operating conditions the concentration and purity of uranium can be markedly improved. The solvent is reconditioned, if necessary, and recycled to the extraction circuit.

Solvent extraction can often yield a better product at less cost in a relatively shorter time.

A good extractant should possess the following characteristics:

Ability to extract uranium from the leach liquor with sufficient selectivity,

Low flammability and low toxicity so that it can be safely handled,

Inexpensive and simple methods for stripping the extracted uranium,

Low solvent loss into the aqueous phase through solubility or incomplete separation of phases after mixing,

Sufficient chemical stability so that it may be recycled, and

Commercial viability

5.1.4.2.1 DAPEX Process

The alkyl phosphoric acids were the first group of reagents for industrial level uranium extraction from sulphuric acid leach liquors. Of these, di-2-ethyl hexyl phosphoric acid (D2EHPA) finds wide use. The process is called the DAPEX process.

The DAPEX process unit was first set up in 1956 at the Ker McGee mill in Shiprock, New Mexico (US). The organic phase consists of a 3–5% (volume/volume) solution of D2EHPA with about the same concentration of tri-butyl phosphate (TBP) in kerosene. The addition of TBP will result in higher extraction through what is known as the *synergic effect*. This effect arises when two reagents cooperate in the extraction of a species in such a way that the extraction is greater than that from a simple summation of the extraction of each reagent acting alone. In the present case TBP, which enhances the extraction by the main solvent, is called a *synergistic agent*.

The extraction of uranium by D2EHPA and other alky phosphoric acids takes place by the mechanism:

$$UO_2SO_4 + 2\ R_2HPO_4 \rightarrow UO_2(R_2PO_4)_2 + H_2SO_4$$

(Aq) (Org) (Org) (Aq)

Stripping of uranium from the scrubbed extract is achieved by shaking with a solution of sodium carbonate.

$$UO_2(R_2PO_4)_2 + 3Na_2CO_3 \rightarrow Na_4[UO_2(CO_3)_3] + 2NaR_2PO_4$$

$$\text{(Org)} \quad \text{(Aq)} \quad \text{(Aq)} \quad \text{(Org)}$$

One of the drawbacks of the DAPEX process is the extraction of ferric iron (Fe^{3+}) also by the solvent. This is minimised by prior reduction of ferric iron to its reduced state (Fe^{2+}) by adding scrap iron. Other metals like vanadium are also extracted to some extent.

Phosphonic acid and phosphinic acid belonging to this group are also used as extractants.

5.1.4.2.2 Extraction by High Molecular Weight Amines

High molecular weight secondary and particularly tertiary alkyl amines extract anionic uranium complexes from sulphate leach liquors. The tertiary amines, which proved to be more selective in the extraction of uranium, soon superseded the alkyl phosphoric acid extractants. West R and Consolidated Mines in South Africa first adopted the amine extraction on a commercial scale in 1957. Many other companies followed suit.

Tertiary amines with aliphatic carbon chains of 10–13 carbon atoms, such as Adogen 363 (Sherex Chemicals) or Alamine 336 (Henkel Chemicals) have been extensively used for the extraction of uranium from sulphuric acid leach liquors. In this process, called the AMEX process, extraction of uranium into the amine phase occurs as the anionic sulphate complex $UO_2(SO_4)_3^{4-}$.

$$2(R_3N^+H)_2SO_4^{2-} + UO_2(SO_4)_3^{4-} \rightarrow (R_3N^+H)_4[UO_2(SO_4)_3^{4-}] + 2SO_4^{2-}$$

$$\text{(Org)} \quad \text{(Aq)} \quad \text{(Org)} \quad \text{(Aq)}$$

Since the extraction occurs through an ion-exchange mechanism, the amine extractants came to be known as *liquid ion-exchangers*.

Usually a 5% (volume/volume) solution of the tertiary amine in kerosene is used for extractions. The extract is scrubbed with dilute sulphuric acid to remove some of the co-extracted impurities like soluble silica, arsenic, vanadium, molybdenum and phosphate. A 5% amine solution can be loaded up to 8–10g uranium per litre. The uranium from the extract is stripped with an acidified sodium chloride, or an alkali carbonate solution.

$$(R_3N^+H)_4[UO_2(SO_4)_3^{4-}] + 4NaCl \rightarrow 4(R_3N^+H)Cl + UO_2(SO_4) + 2Na_2SO_4$$

$$\text{(Org)} \quad \text{(Aq)} \quad \text{(Org)} \quad \text{(Aq)} \quad \text{(Aq)}$$

A single stage of solvent extraction results in only partial recovery of uranium from the aqueous phase. For achieving greater extraction efficiency, a multiple extraction process using the counter-current technique is employed. The process involves normally 3–4 extraction stages, 2–3 scrubbing stages and 2–4 stripping stages. Different types of equipment have been developed for the counter-current extraction.

For producing a high grade uranium product suitable for conversion to uranium hexafluoride (for isotopic enrichment), some mills adopt a combination of ion-exchange and solvent extraction. Some examples are "ELUEX" and "BUFFLEX" processes. They consist of,

- Ion-exchange adsorption of uranium from sulphuric acid leach solutions,

- Elution of uranium from the column with 10% sulphuric acid,

- Amine extraction of uranium from the eluate,

- Stripping of uranium from the extract by ammonium sulphate – ammonia mixture of pH 3.5–5.0,

- Precipitation of uranium with ammonia, and

- Calcination to oxide for conversion to fluoride.

5.1.5 Yellow Cake

Yellow Cake is a rather loosely used term in the uranium industry. Historically, the yellow-coloured concentrate, containing 70% or more of uranium, recovered through a series of chemical precipitations after the ore is leached with sulphuric acid or sodium carbonate, is called by this name. With improvements in technology, some mills started producing calcined uranium products that were dirty green and black in colour with compositions close to U_3O_8. They are also referred to as Yellow Cake in the industry. The production methods of these end-products depend on the steps through which the leach liquors pass.

When uranium is leached from the ore by carbonate leaching, the product solution goes directly for precipitation by sodium hydroxide. Alternatively, the carbonate is destroyed by addition of acid and the uranium precipitated with ammonium hydroxide, magnesia or hydrogen peroxide. The resulting cake is filtered, washed, dried and packed as Yellow Cake.

When ion-exchange method is used to separate and concentrate uranium from sulphuric acid leach liquors, the chloride or nitrate eluate that goes for uranium precipitation contains 8–10g U/l and varying amounts of impurities like iron (ferric), vanadium, phosphate and sulphate. If a base like sodium hydroxide, ammonium hydroxide, or magnesia is added the precipitated product carries some or all of the impurities along with the uranium. To improve the quality of the product, precipitation is generally carried out in two stages. In the first stage, slaked lime (calcium hydroxide) is added till the solution attains a pH of about 3.5. At this stage, most of the impurities are precipitated along with a small fraction of uranium. This cake, called *iron-gypsum* cake, is filtered, washed and transferred to the leaching circuit for recovery of uranium. To the filtrate a base like sodium hydroxide, ammonium hydroxide or magnesia is added to precipitate uranium. The resulting cake is filtered, washed and dried.

5.1.6 Production of Yellow Cake in India

Laboratory studies on the extraction of uranium from the ore began in 1953 at BARC. By 1958, a set of optimum conditions for the process were worked out. Taking into consideration the lower uranium content of the ore (about 0.04% U) and the behaviour of associated minerals like apatite (calcium phosphate) in the leaching stage, leaching with sulphuric acid at a controlled pH of 1.5–1.8 is adopted. Pyrolusite, a naturally occurring manganese dioxide mineral, is used as an oxidant. The Jaduguda (Jharkhand) mill with a capacity of 1,000 tonnes of ore per day was fully commissioned by 1968. The extraction process was developed in-house by BARC scientists and engineers. When the mill was about to be commissioned, the Uranium Corporation of India Ltd (UCIL), constituted in 1967 as a public sector undertaking under the Department of Atomic Energy was entrusted with uranium mining and milling operations in the country. The process adopted in Jaduguda Uranium Mill generally follows the standard acid leach–ion-exchange recovery scheme. The Yellow Cake is then sent to Nuclear Fuel Complex, Hyderabad, for further refining and fabrication of nuclear reactor fuel bundles.

5.2 Uranium Production in India

UCIL and DAE do not, as a matter of policy, officially publish data on grades of ores processed in their mills or the annual output. However, some international organisations project values in their publications. According to WISE (World Information Service on Energy) Uranium Project reports, a vein type deposit containing 0.042–0.051% uranium is being mined in Jaduguda. The ores in other areas may be assumed to be of about the same grade.

To meet the growing uranium demand for nuclear power generation in the country, the UCIL has expanded its mining and milling capacities. The Bhatinda mine, near Jaduguda, started production in 1986–87, supplying 150 tonnes of ore per day to the Jaduguda mill. In 1995, another mine was developed at Narwapahar, 12 km from Jaduguda. This ore is processed in Jaduguda as well. At present, the capacity of the Jaduguda mill stands at about 2,010 tonnes of ore per day. A new uranium mill to process about 3,000 tonnes of ore per day was commissioned in 2007 at Turamdih, 24 km from Jaduguda. It is also expected to process the ore from nearby mines like Turamdih, Mohuldih, Banduhurang, and Bagjata. On this basis, the Turamdih complex, when it reaches full production, could add 250–300 tonnes of uranium per year. A new mine and processing plant were commissioned at Thummalapalle in Andhra Pradesh. The mine will be completely underground at a depth of 300m.

The first processing plant, constructed next to the mine, has the capacity to process 3,000 tonnes of low-grade ore, a uranium content of 0.2%. It is claimed that when they become fully operational, the Tummalapalle mines will cater to the requirements of 25% of the nuclear power plants in the country.

Other deposits under study are Gogi (Karnataka), Lambapur-Peddagattu (Nalgonda, AP). Nearly twenty thousand tonnes of uranium ore were found in West Khasi Hills in Meghalaya. But the civil society is opposed to mining operations. A new deposit containing an estimated 5,185 tonnes was found recently at Rohil in Rajasthan's Sikar district.

The extraction of uranium from the low-grade ores involves processing of large quantities of ore resulting in the generation of large volumes of solid waste and effluent. With greater public awareness of health hazards and stringent environmental guidelines, the management of this waste has become a major issue. In addition, the production of uranium from these low-grade ores is expected to result in higher production costs of uranium when compared to the international price. These factors naturally push up the cost of nuclear power.

The World Nuclear Association's August 2012 estimates say that India produced 209 tonnes of uranium in 2009. The annual production rose to 400 tonnes in 2010 and 2011. With an ambitious programme for nuclear power generation, India's additional annual uranium requirement is expected to be 1,500 to 2,000 tonnes, for which it has to depend on foreign supplies.

To meet the rising requirements of uranium for its nuclear power programme, India is also planning, in the wake of the recent relaxation of international sanctions, to set up a new company to acquire uranium mines in South Africa, Kazakhstan, and Mongolia.

5.3 Refining and Conversion

The yellow cake from a uranium mill requires further processing before it can be turned into reactor fuel through refining. This involves the removal of impurities like boron, lanthanides, cadmium etc., which strongly absorb neutrons to extremely low levels. The most effective method adopted to achieve this level of purity, called nuclear purity, is solvent-extraction.

The yellow cake is dissolved in nitric acid and the insoluble residue is removed by filtration or centrifugation. The uranium in the solution is then selectively extracted into an immiscible organic solvent, tri-butyl phosphate (TBP) diluted with a hydrocarbon solvent leaving practically all the impurities in the aqueous phase. The impurity level in the uranium-bearing organic phase is further reduced to the desired levels by washing it with dilute nitric acid. The separated organic phase is then washed with water when pure uranyl nitrate is transferred back into the aqueous phase. The purified uranium solution goes to the next stage, called conversion. Addition of ammonia to the uranium-bearing solution results in the precipitation of yellow ammonium diuranate. This is filtered, washed and heated to obtain uranium oxide (UO_3). Further processing of this material depends on the type of fuel. The fuels are,

- Natural Uranium Oxide
- Natural Uranium metal

- Enriched Uranium Oxide
- Mixed Oxides of Uranium and Plutonium (MOX)

5.3.1 Natural Uranium Oxide Fuel

Presently CANDU (CANada Deuterium Uranium) reactors and PHWRs (Pressurised Heavy Water Reactors) in India are fuelled by natural uranium dioxide (UO_2). The refined uranium oxide obtained from the refinery is reduced to UO_2 in a rotary kiln at 700° C while passing a mixture of hydrogen and nitrogen.

$$UO_3 + H_2 \rightarrow UO_2 + H_2O$$

5.3.2 Natural Uranium Metal Fuel

Power reactors using uranium metal as fuel are mainly the MAGNOX reactors in the UK. In India, the research reactors CIRUS (now shut down) and DHRUVA are operated with uranium metal as fuel. The Uranium Metal Plant (UMP), which is located at BARC, Trombay (Mumbai) produces all the metal required for these reactors. The process for making metal described below is based on Indian experience.

Uranium dioxide is first converted into uranium tetrafluoride (UF4) by heating in a rotary furnace while passing dry hydrogen fluoride (HF) gas in a counter-current mode. The UF4, a green powder, is then mixed with magnesium chips and heated in a steel reactor lined with magnesium fluoride powder, to give uranium metal. The reaction is highly exothermic. The uranium metal, which is in a molten condition when formed, collects in a cylindrical mould at the bottom of the reactor. After cooling, the metal ingot is separated from the magnesium fluoride slag. At present, each ingot produced in India weighs about 450kg. These ingots are transferred to the Fuel Fabrication Plant, also located at Trombay, for fabrication as fuel elements for the reactors.

5.3.3 Enriched Uranium Oxide

Most of the power reactors operating in the world are Light Water Reactors (LWR) using low enriched uranium dioxide as the fuel. The production of enriched uranium oxide for the purpose consists of four stages. In the first stage, natural uranium tetrafluoride (UF_4) is produced as described above. In the second stage, UF_4 is converted to uranium hexafluoride (UF_6) (also called 'Hex') by feeding it along with gaseous fluorine to a fluidised-bed reactor. The UF_6 product, which comes out as a gas, is condensed into a solid and stored in steel cylinders. In the solid form, UF_6, is a dense and, crystalline resembling rock salt. UF_6 does not react with oxygen, nitrogen or carbon dioxide, but reacts vigorously with water or water vapour liberating the highly corrosive hydrogen fluoride. For this reason, it is always handled in leak-proof containers and processing equipment. Only Ni or Al or their alloys are suitable for handling UF_6. This chemical also attacks most of the greases and lubricants that are generally used in the chemical industry. As a part of the Manhattan Project, special lubricants, mostly fluorinated hydrocarbons, were developed for use in systems handling UF_6. In spite of its corrosive properties, UF_6 is preferred for isotopic enrichment because of its unique properties. It can be conveniently converted into gaseous form for processing, liquid form for filling and to a solid form for storage, all at temperatures and pressures commonly used in chemical processes (Room temperatures to 70°C and 1–1.5 atmosphere pressure). The complex technology of isotopic enrichment of uranium is discussed in the following chapter.

For fabrication of the enriched uranium fuel, the solid enriched UF_6 contained in the cylinders is changed to gaseous state by heating. It is then allowed to react with a solution of ammonia forming ammonium diuranate solid. The product is then converted to the oxide form and pressed into pellets.

5.3.4 Mixed Oxide (MOX) Fuel

MOX fuel is a mixture of plutonium and uranium oxides. Part of the Light Enriched Uranium (LEU) fuel used in LWRs can be replaced by MOX fuel, without making design changes in the reactor. The plutonium oxide content may vary from 1.5 wt% to 25–30 wt% depending on the type of the reactor. MOX fuel is used to a limited extent in nuclear reactors in Japan, Switzerland, Germany, Belgium, and France. It is also used to some extent in India's Tarapur thermal reactors. Efficient burning of the plutonium can, however, be achieved only in fast reactors. The MOX fuel is a way of utilising surplus weapons-grade plutonium available from the nuclear weapons being dismantled. It is also an alternative for safe storage to prevent its theft and proliferation.

5.4 Fuel Fabrication

Fuel fabrication represents the final stage of the front-end of the nuclear fuel cycle. Fuel assemblies are precision-engineered products conforming to the physical characteristics of a specific reactor.

5.4.1 Fuel for Light Water and Heavy Water Reactors

A majority of power reactors presently operating in the world are in the categories of PWR, BWR and PHWR. They use uranium in the form of oxide (UO_2) as fuel.

For light water or heavy water reactors the fuel fabrication process starts with pressing uranium oxide (low-enriched or natural) into cylindrical form (about 1 cm diameter x 1.5 cm length) and sintering at 1600–1700°C yielding dense ceramic pellets (Fig.1) with low porosity. These pellets are centreless ground and loaded into Zircaloy tubes. The gap between the fuel and cladding is filled with helium gas to improve heat conduction. The tubes are sealed at both ends and bundled together into fuel assemblies (fuel bundles) (Fig.2). The number of assemblies in a reactor core depends on the core size and reactor power output. A typical 1,100MWe PWR contains 193 fuel assemblies composed of 51,000 fuel rods four metres in length, loaded with approximately 18,000,000 fuel pellets. The BWRs carry 74–100 fuel rods per assembly and about 800 assemblies in the core. The CANDU and PHWR fuel bundles are short, half a metre in length. Each bundle carries 19–34 fuel rods. Up to 12 bundles lie end-to-end in a fuel channel. A typical core loading in a CANDU reactor is of the order of 4,500 bundles.

Fig. 1: Uranium Oxide Pellets

Fig. 2: CANDU Fuel Bundles

5.4.2 Metal Fuel

Uranium metal is produced as angot by the reduction of uranium tetra fluoride (UF_4) with magnesium in a steel pressure vessel lined with magnesium fluoride. The ingot is induction melted and cast into billets, which are then hot rolled to rod shape (about 2.5cm. diameter), heat treated, straightened and machined to the required dimensions. In India, the fuel rod is canned in an aluminum tube, which is then sealed by welding aluminium plugs at either end.

5.4.3 MOX Fuel

The procedure for fabrication of MOX fuel is similar to that of the uranium oxide fuel. But in view of the extreme toxicity and high level of radioactivity of plutonium, the entire operation is carried out taking special precautions.

5.5 Global Facilities for Refining, Conversion and Fuel fabrication

Major nuclear powers like US, Russia, France, UK, and Canada have their dedicated domestic facilities for refining, conversion, and fabrication of nuclear fuel (Tables 1–4). Some of these countries import yellow cake for this purpose. These facilities are big enough to cater to their own needs as well as meet the requirements of other countries.

Fuel fabrication is an important component of the fuel cycle. It can influence fuel cycle costs. The worldwide requirement of LWR fuel was about 7,500 tHM (tonnes of heavy metal- natural or LEU or MOX) in 2007. This requirement is expected to increase to 9,700 tHM by 2015. But the present capacity is close to 13,000 tHM per annum, which is in excess of the demand. The fuel requirement for CANDUs (PHWRs) is about 3,000 tHM/year and for gas-cooled reactors about 400 tHM/year. It appears that enough capacity exists to meet the demand till 2020.

Table 1: World Primary Uranium Hexafluoride Conversion Facilities

Company and Location Capacity (t/y)
Cameco, Ontario, Canada 10,500
Westinghouse, Springfield, UK 6,000
Rosatom, Russia 24,000
Comurhex, (Areva)' Pierrelette, France 14,350
Converdyn, Metropolos, US 17,600
CNNC, Lanzhou, China 3,000
IPEN, Sao Paulo, Brazil 90
AEOI, Isfahan, Iran 193
Total 75,733

Source: WISE uranium Project, Updated April 13, 2013.

Table 2: World LWR and BWR Fuel Fabrication Capacity(t/y)

Country	Conversion	Pelletising	Rod/Assembly
Belgium	0	700	700
Brazil	160	160	280
China	400	400	450
France	1,800	1,400	1,400
Germany	800	650	650
India	48	48	48
Japan	475	1,800	1,724
Kazakhstan	2,000	2,000	0
Korea	600	600	600
Russia	1,650	1,400	520
Spain	0	300	300
Sweden	600	600	600
UK	2,150	1,800	2,060
US	3,900	3,900	3,450

Source: World Nuclear Association September 2011.

Table 3: World PHWR Fuel Fabrication Capacity (t/y)

Country	Rod/Assembly
Argentina	160
Canada	2,700
China	200
India	435
Pakistan	20
Korea	400
Romania	240

Source: World Nuclear Association, September 2011.

Table 4: World MOX Fuel Fabrication Capacities (t/y)

Country and location	2009	2015
France, Melox	195	195
Japan, Tokai	10	10
Japan, Rokkasho	0	150
Russia, Mayak, Ozask	5	5
Russia, Zheleznogorsk	0	60(?)
UK, Sellafield	40	0

Source: World Nuclear Association, May 2012.

(Note: The U S is building a fuel fabrication plant in Savannah for utilising the plutonium released from nuclear weapons. It is expected to start production in 2018).

Information on the enrichment facilities is given in the next chapter.

5.6 Refining and Fuel Fabrication Facilities in India

The reactors set up in India so far are PHWRs (except two at Tarapur). The first plant to go into production in India for the refinement of yellow cake and its conversion to uranium metal ingots was commissioned in Trombay, Mumbai at the Bhabha Atomic Research Centre in 1959. It was originally designed to produce 25 tonnes of the metal per year for providing fuel to the 40MWt CIRUS research reactor. The capacity was later enhanced to meet the fuel requirements of the 100MWt DHRUVA reactor for research and production of plutonium. The uranium metal fuel for these two reactors is fabricated at the Fuel Fabrication Plant in Trombay.

With more power reactors coming on stream, the Nuclear Fuel Complex (NFC) was set up at Hyderabad in 1971 for refining, conversion and fuel fabrication based on the know-how developed at BARC. Currently, the NFC supplies the fuel assemblies for all PHWRs. In addition, the NFC fabricates the LEU fuel assemblies for the two BWRs at Tarapur using imported LEU hexafluoride.

In addition to uranium oxide production, NFC has facilities for the production of zirconium metal (low in hafnium content) and its alloys (Zircaloy). The products include components such as fuel tubes, coolant tubes, calandria tubes, spacer pads, end-plugs, and end-plates. Equipped with various chemical, metallurgical and mechanical operations, the NFC is a truly integrated facility for nuclear fuel fabrication.

As a research centre, the BARC has also developed and fabricated a wide variety of nuclear fuels required for experimental purposes and test reactors. MOX fuel bundles for the test in the BWR at Tarapur were first fabricated here. Based on this know-how, an advanced fuel fabrication facility has been set up at Tarapur for the production of MOX fuel. Uranium-plutonium carbide fuel pins for the Fast Breeder Test Reactor (FBTR) at Kalpakkam are also fabricated at BARC.

In the wake of relaxation of sanctions, efforts are on to import light water reactors from Russia, France, and the US. In the initial stage, the LEU for these reactors is to be supplied by the reactor vendors. India will be able to meet fuel requirements for these reactors indigenously in due time.

6

Uranium-235 Enrichment

6.1 Natural Uranium

Natural uranium is always of the same isotopic composition: 99.27% U-238, 0.72% U-235 and traces (0.0054%) of U-234. At the time, the solar system with Earth as its constituent was created nearly five billion years ago the two main isotopes were believed to be present in about equal concentrations. With the comparatively rapid decay rate of U-235 (half-life 7.1×10^8 years) as compared to that of U-238 (half-life 4.5×10^9 years) the former has reached the present low level. The time taken for this decay process can be calculated as about 6 billion years. Significantly enough, this value approximately corresponds to the age of Earth.

U-235 is the only natural isotope that can be used for direct nuclear energy production in a reactor. Stated briefly, the basic requirement of a nuclear reactor is an assembly of fissionable material (U-235) of a critical size that can release energy through a sustained fission chain reaction. The construction of the first nuclear reactor by Enrico Fermi at Chicago in 1942 is described earlier.

All light water moderated reactors require critical assemblies of uranium containing U-235 enriched to levels of 3% or more. For assembling an atom bomb the enrichment should be at least 80–90%.

6.2 The Oklo Natural Nuclear Reactor

At this point, we will digress and learn about an extraordinary event arising from the presence of a higher concentration of U-235 in natural uranium in the early times. This event occurred about 1.8 billion years ago at a place called Oklo in Gabon, Equatorial Africa. The uranium mine located at this place was supplying uranium to the nuclear industry. During 1970, the analysis of a shipment of uranium showed that its U-235 content was much smaller than the normal value of 0.72%. This was unusual because the value was considered constant. This very anomalous phenomenon was investigated carefully by scientists who came to the conclusion that this was the result of an extraordinary event.

According to them, about two billion years ago the proportion of uranium-235 in natural uranium has been higher at around 3%. During this period Earth's atmosphere contained very little oxygen and uranium was present in the elemental form and insoluble in water. When enough oxygen accumulated in the atmosphere during the Proterozoic era (about 2.5 billion to 5 million years ago) and entered the groundwater, the uranium in the rocks began to interact with oxygen and form the water-soluble uranyl ion (UO_2^{2+}) species. The uranyl species then entered the water streams, which flowed into an algal mat at Oklo. The microorganisms present in the algal mat had the unique capacity to collect and concentrate this uranium. Eventually, the accumulated uranium containing the fissionable U-235 (with 3% abundance) exceeded the critical mass. The water acted as a neutron moderator and initiated a nuclear chain reaction. This reaction was, however, not explosive but was self-regulating. With the rise of heat output, the water boiled away and the nuclear reaction slowed down. It started again when sufficient water accumulated. The reaction thus ran gently at a power level of the order of 100 kW and in the process used up a significant part of U-235. There were at least a dozen of these reactors in the area putting out energy equivalent to 100 megaton bombs until their U-235 content in the natural uranium deposit was depleted to the observed low levels. According to George Cowan, a former Manhattan Project scientist, "*After the reactor has shut down, the evidence of its activity was preserved virtually undisturbed through the succeeding ages of geological activity.*" He further said, "*In the design of fission reactors, man was not an innovator, but an unwitting imitator of nature.*" The Oklo phenomenon is a unique example of the biogeochemical processes by which minerals are separated and concentrated. At Oklo, these processes helped creating conditions for a natural nuclear reactor to become operational. Fortunately, a recurrence of such an event now is just not possible because at the present concentration levels of U-235 (0.72%) the quantity of natural uranium with ordinary water as moderator, which must be assembled for achieving criticality for starting nuclear chain reaction is almost infinite.

6.3 Manhattan Project

As mentioned earlier, the Manhattan Project was launched by the US in 1942. The most complicated part of this programme was the production of ample quantities of uranium enriched to 80–90% with respect to U-235 for assembling the bomb. This required development of processes for separating the U-235 from the abundant U-238. Since isotopes of an element are chemically identical, no chemical method could be used for the purpose. The only alternative was through methods exploiting the slightly different physical profiles arising from the small difference in their isotopic masses. Three methods were pursued at the Manhattan Project: Electromagnetic Method (Code Name Y-12), Gaseous Diffusion Method (Code Name K-25) and Thermal Diffusion Method (Code Name S-50). Centrifugation method was also considered but abandoned due to the non-availability of technology for operating the high-speed rotors for large-scale separation.

Three plants were built at Oak Ridge, Tennessee, for making highly-enriched uranium (HEU) containing 80–90% uranium-235, which was given the code-name Oralloy (Oak Ridge Alloy). The first stage was S-50 in which the enrichment level of 1–6% was achieved through thermal diffusion technique. This material was fed into the gaseous diffusion plant (K-25), which yielded a product enriched to about 25% U-235. This product was finally fed into the electromagnetic separation system (Y-12), which boosted the U-235 content to >84%. A critical assembly containing 64 kg of HEU went into the atomic bomb that was dropped over Hiroshima on August 6, 1945. Currently, there are about 2,000 tonnes of HEU in the world produced mostly for nuclear weapons, naval propulsion, and other special purposes.

6.4 Enrichment Methods

Uranium enriched to 20% or more U-235 is called *Highly Enriched Uranium* (HEU). Uranium enriched to more than natural abundance but to less than 20% is called *Low Enriched Uranium* (LEU). Nuclear weapons require 80–90% enrichment levels while nuclear reactors usually require 3–5% enrichment level. An enrichment plant takes feed material with a certain enrichment and yields a product with higher enrichment, generating a waste or tails of lower enrichment. Isotope enrichment methods can be classified as:

- Enrichment by Thermal Diffusion
- Enrichment by Gaseous Diffusion
- Enrichment by Electromagnetic Method
- Enrichment by Aerodynamic Processes
- Enrichment by Centrifugation
- Enrichment by Laser Technique.

6.4.1 Separation Factor

The degree of separation that can be achieved in a mixture containing two species is measured by a parameter called separation factor. In the present context, it represents the ratio of two isotopes after processing divided by their abundance ratio before processing. This will now be explained in terms of the gas diffusion method, the first successful method for the enrichment of isotopes. The method exploits the different rates of diffusion of gases through a porous barrier.

According to Graham's law, the rate of diffusion of a gas is inversely proportional to the square root of its density. This is due to the fact that, at the same temperature, the molecules of a light gas move on the average with a higher speed than those of a heavier gas. Since the density of a gaseous compound is also directly related to its molecular weight, a gas of lower molecular weight will pass through a porous barrier faster than one of higher molecular weight. In a mixture of two isotopic molecules, those containing the lighter isotopes will diffuse more rapidly. If M_1 is the molecular weight of the form containing the lighter isotope, and M_2 that of the form with the heavier isotope, the ideal separation factor (s) for diffusion will be represented by $\sqrt{M_2/M_1}$. When the separation factor is very small (i.e. slightly greater than unity) only a small enrichment can be achieved in single diffusion stage. To achieve the required enrichment, the process material should therefore be subjected through a series of stages, a process called a cascade process. The greater the enrichment required the larger are the number of stages needed (see later).

6.4.2 Separative Work Unit (SWU)

The *Separative Work Unit* (SWU) used in uranium enrichment processes is a complex unit. It is a function of the amount of natural uranium processed and the degree to which it is enriched (i.e. the extent of increase in the concentration of U-235 isotope relative to the remainder) and the level of depletion in the remainder (tails). Using a formula derived from thermodynamical considerations, it is possible to calculate the number of separative work units, expressed as kg-SWU, required to produce a kilogram of uranium enriched to any specified concentration of U-235, given the starting concentration in the material and the concentration left in the tails. The number of SWUs required during enrichment increases with decreasing levels of U-235 aimed in the tails stream. It is also indicative of the energy used in the enrichment operation. The SWU concept will now be illustrated with a few examples.

Production of one kilogram of HEU with 93% U-235, starting from 226 kilograms of natural uranium(containing 0.72% U-235 and leaving behind 225 kilograms of tails containing 0.3% U-235), requires 200SWU. Thus, the enrichment requirement for a nuclear weapon containing 60 kilograms of HEU would be 60kg HEU x 200SWU per kilogram of HEU which works out to 12,000 SWU.

Producing one kilogram of LEU with 5% U-235 starting from 11.5 kilograms of natural uranium, leaving behind 10.5 kilograms of tails containing 0.3% U-235, requires 7.2 SWU. Thus, the annual separative work unit requirement for a 1,000 MW light water reactor is 20,000 kg of LEU x 7.2 SWU per kilogram or 144,000 SWU.

The power requirements for uranium-enrichment plants differ with the technology used. They range from 100–150 kWh per SWU in a centrifuge plant to 2,000–3000 kWh per SWU in a gaseous diffusion plant and about 4000 kWh per SWU for aerodynamic technologies. Laser enrichment technologies require 100–200 kWh per SWU.

The power requirements for uranium-enrichment plants differ with the technology used. They range from 100–150 kWh per SWU in a centrifuge plant to 2,000–3000 kWh per SWU in a gaseous diffusion plant and about 4000 kWh per SWU for aerodynamic technologies. Laser enrichment technologies require 100–200 kWh per SWU.

For producing 60 kilograms of 93% U-235 HEU for assembling a nuclear weapon, the power requirement for the gaseous diffusion method will be 12,000 SWU x 2,500 kWh per SWU or 30,000,000 kWh. Making about a dozen such bombs per year requires the full output of a 50 MW power plant. In contrast, a gas centrifuge plant would require just 2.5 MW of dedicated power supply for making an equal number of weapons per year!

6.4.3 Feedstock for Enrichment

Practically all isotope separation methods excepting the electromagnetic method require a chemical compound that maintains the small differences in the isotopic masses of uranium. Further, it should also be amenable for easy conversion to gaseous or liquid form. The most suitable uranium molecule that satisfies these conditions is uranium hexafluoride (UF_6, also called 'hex'). With fluorine being monoisotopic $^{235}UF_6$ and $^{238}UF_6$ molecules have masses of app. 349 and 352 amu respectively. Uranium hexafluoride may thus be considered as a mixture of two isotopic forms. The rate of diffusion of these molecules is therefore determined by the uranium isotopes only and the separation factor for single diffusion stage is given as $\sqrt{352/349}= 1.0043$.

Preparation of uranium hexafluoride is described in an earlier chapter. It is a white solid, which sublimes to gaseous state when heated to 56⁰ C. Plants commercially producing UF_6 are located in US, Canada, France, UK, Russia, China and Brazil with a total operating capacity of 76,090 tonnes per year (WNA Market Report 2009).

For electromagnetic separations the feed material for ion generation is uranium tetrachloride (UCl_4). It is produced by the reaction of carbon tetrachloride with uranium dioxide at about 370⁰C. Its main advantage is that it is not as corrosive as UF_6.

6.4.4 Thermal Diffusion Method

The thermal diffusion method is based on the principle that when a mixture of isotopes in gaseous or liquid form is placed in a vessel, part of which is hotter than the remainder; the lighter molecules tend to concentrate in the regions of higher temperature through convection and diffusion processes. The thermal diffusion enrichment plant (S-50) set up at Oak Ridge started operating in September 1944. It consisted of 2100 columns each approximately 15m long. Each of these columns consisted of three concentric tubes. Cooling water was circulated between the outer and middle tubes while the inner tube carried steam. The annular space between the inner and middle tubes was filled with liquid UF_6. Cooling this vertical film of UF_6 on one side and heating on the other produces convection currents causing an upward flow on the hot surface and a downward flow along the cooler side. Under these conditions, the lighter $^{235}UF_6$ molecules diffuse towards the warmer surface and the heavier $^{238}UF_6$ molecules to the cooler side. This combined convection-diffusion effect leads to the concentration of the lighter ^{235}U at the top of the film while the heavier ^{238}U goes to the bottom (Fig.1). This plant provided enriched uranium for further enrichment in the gaseous diffusion and electromagnetic separation plants. The plant was shut down after one year as the operation involved very high power consumption.

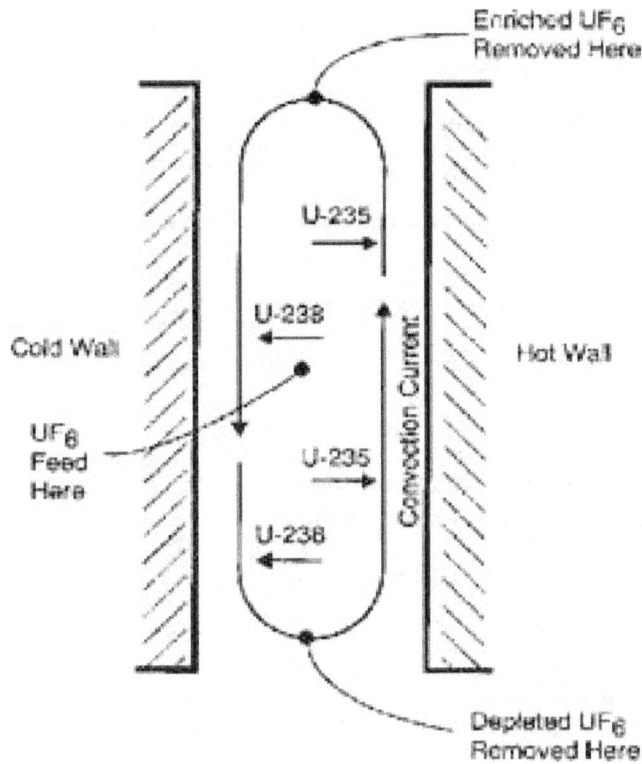

Fig. 1: U-235 Enrichment by Thermal Diffusion

6.4.5 Gaseous Diffusion Method

The gaseous diffusion method (explained briefly earlier) is based on the fact that at thermal equilibrium two isotopes with the same energy will have slightly different average velocities. Thus, if a gas consisting of two isotopes is allowed to diffuse through a membrane into an evacuated vessel, the molecules containing lighter isotope diffuse faster than the molecules containing the heavier isotope. The ideal separation factor for $^{238}UF_6$ and $^{235}UF_6$ as shown earlier is 1.0043 for one diffusion stage.

This ideal value can be almost realised only at the beginning of the diffusion process because as the diffusion proceeds there is a tendency for the light component to diffuse back. This will result in the decrease of the separation factor. Further, when diffusion is allowed to continue long enough for equilibrium conditions to be established, the concentrations on both sides of the barrier tend to become identical. For this reason, there has to be a compromise between the fraction of the gas permitted to diffuse and the resultant separation factor. This is achieved by allowing only about half of the gas to pass through the barrier.

For obtaining true diffusion (or effusion), the choice of an appropriate barrier is very crucial. It is necessary that holes be less than one-tenth the mean free path of the molecules. According to calculations, the pores in the barrier would have to be about one-millionth of a centimeter in diameter at ordinary pressures. As the pressure decreases there would be an increase in the mean free path of the molecules and consequently, an increase in the pore size is permissible. The barrier should also be compatible with the highly corrosive UF_6, should be thin enough to provide minimum resistance to the gas flow, and mechanically strong enough to withstand the pressure differential. One known method for making such a barrier is etching a thin sheet of silver-zinc alloy by means of hydrochloric acid. The acid would dissolve some of the zinc atoms, leaving a large number of submicroscopic holes in the sheet of metal. Barriers made of sintered nickel powder have also been found suitable. However, the technology of making barriers is classified.

Under experimental conditions, the separation factor in a single stage is lower with a value of 1.0014 as compared to the ideal value of 1.0043. This is due to factors like barrier efficiency, back pressure efficiency, and mixing efficiency. Since the proportion of U-235 is thus, raised by a factor of only 1.0014 in each diffusion stage, it is necessary to carry out the diffusion process through a large number of stages to obtain the desired enrichment. For n successive stages the overall separation factor will be $(1.0014)^n$. Thus, for a ten-fold (7.2%) enrichment, about 1800 stages are required and for 99% enrichment about 4,000 stages. For achieving the required degree of separation the principle of cascade separation involving a device comprising a large number of interconnected stages is adopted. In this device, the enriched part from any one stage becomes the feed for the next stage while the depleted material, which still contains a considerable percentage of the desired isotope, is mixed with the raw material of the preceding stage. The plant is designed to make the flow from stage to stage automatic and continuous (Fig. 3).

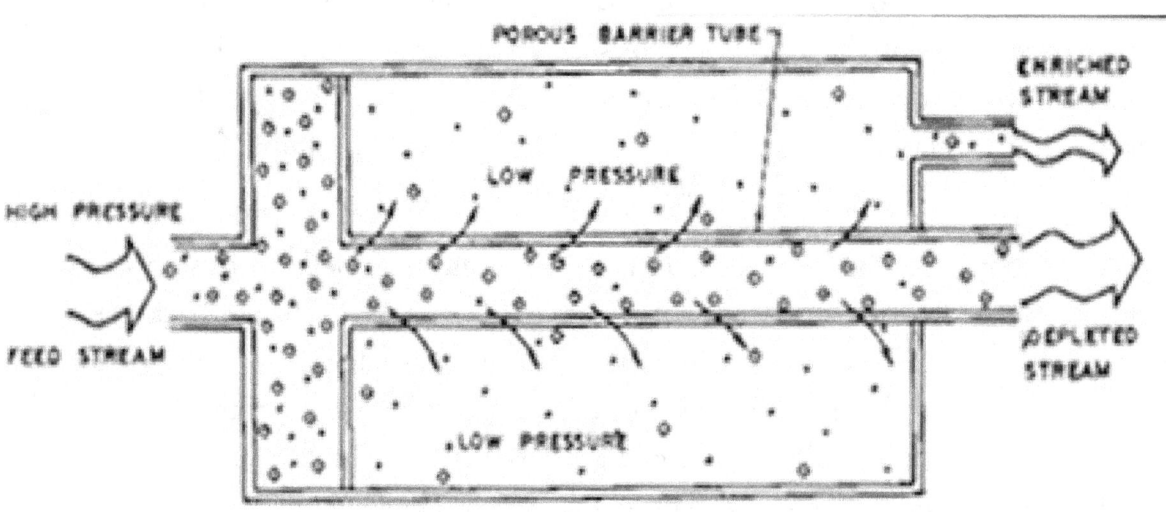

Fig. 2: Gaseous diffusion chamber

Uranium-235 Enrichment | 73

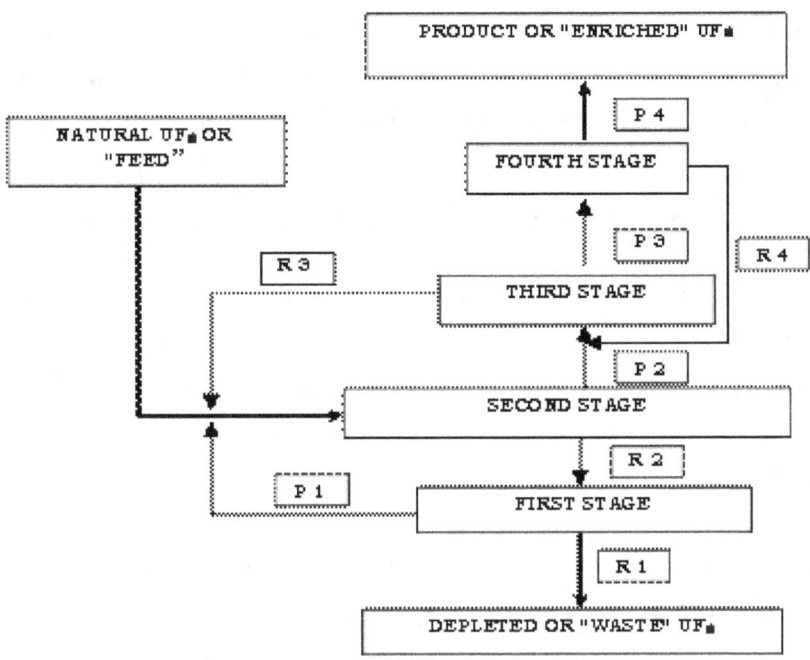

Fig. 3: Cascade Separation (schematic)

The construction of the first gaseous diffusion plant for uranium enrichment (K-25) began at Oak Ridge in the summer of 1943. It took nearly two years to make the plant operational. The diffusion plant building at Oak Ridge and another building constructed later at Paducah covered about 420 acres of land. More than 400 million gallons of water, enough to serve a city of two million people were circulated every 24 hours. Electrical consumption for these two units was about twice that of New York City at the time. With thousands of miles of piping and hundreds of acres of diffusion barriers, the enormous structures (Fig.4) are considered a tribute to "the courage, persistence, as well as the scientific and technical ability" of the physicists, chemists, and engineers.

Fig. 4: U-235 Enrichment plant by gaseous diffusion at Oak Ridge

The highly energy-intensive gaseous diffusion process developed by the US produced HEU during WWII and commercial power reactor grade LEU later. Since the 1960s, the US facilities have been used primarily to produce commercial LEU with the last remaining HEU facility shut down in 1992. China and France currently have gaseous diffusion plants in operation. Russia's enrichment facilities have been converted from diffusion to centrifuge technologies. UK's gas diffusion facility was shut down and dismantled.

In 1979, more than 98% of the global demand for enriched uranium was met through the gaseous diffusion process. This fell to 25% in 2007. By 2017, all the diffusion plants will be shut down making space for the more energy efficient advanced gas centrifuge technology.

6.4.6 Electromagnetic Method

Enrichment by the electromagnetic method, which is essentially based on the principle of the mass spectrograph, was originally undertaken for meeting the need for pure U-235 for experimental purposes. Success in these efforts led to the development of the Calutron by E.O. Lawrence and co-workers for preparing larger quantities of enriched uranium. Calutron is an abbreviation for "California University Cyclotron" because it made use of the magnet from one of University of California Cyclotrons. The project was operated at Oak Ridge under the code name Y-12.

In the Calutron, a positively-charged particle beam will follow a circular path when passing through a uniform magnetic field. The exact path of the positively-charged particles depends on their masses. For a given electric and magnetic field, the square of the radius of the path is proportional to the mass of the particle. The isotopes of a given element can be collected by placing collecting pockets in the appropriate positions. The $^{235}U^+$ and $^{238}U^+$ ions which have the same charge and kinetic energy but different masses will have slightly different paths when moving through a magnetic field. These can be separated and collected. The uranium ions are generated from UCl_4 vapour by bombardment with electrons (Fig.5).

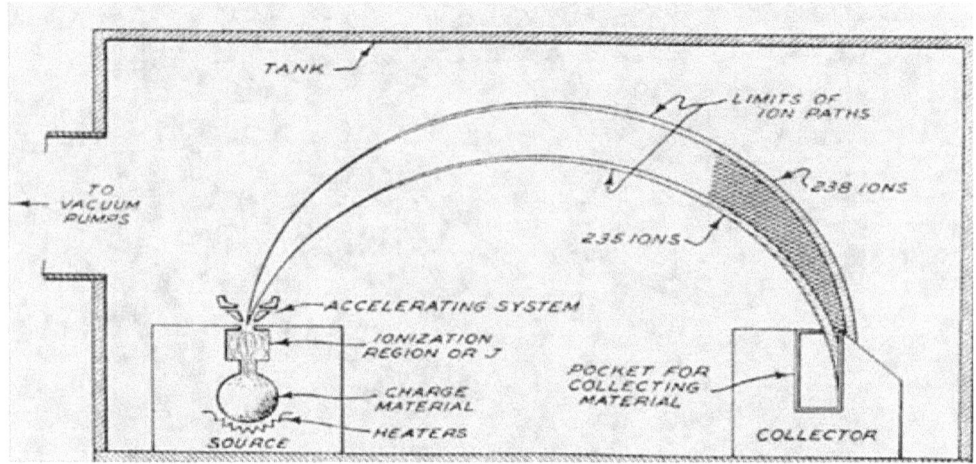

Fig. 5: Calutron

Because the electromagnetic method is a batch process, with each unit requiring a long time to produce a small amount of U-235, many Calutron units had to be installed for preparing HEU in the required quantity. As sufficient copper was not available for the construction of the gigantic magnet component of the battery of Calutrons, silver loaned from the US Government's silver reserves of 14,700 tonnes was used for the purpose!

Two sets of Calutrons were used for enrichment. The first set, called Alpha Calutrons, enriched the 8% U-235 from the gaseous diffusion plant to between 12 and 20% U-235. The second set, called Beta Calutrons, took the Alpha product and enriched it into weapons-grade material.

The electromagnetic method was abandoned because of high power consumption. Iraq adopted this method for making enriched uranium for its aborted nuclear weapons project because of its relative simplicity and Iraq's ability to procure the materials required for assembling the device without encountering technology transfer obstacles.

6.4.7 Aerodynamic Methods

Aerodynamic separation methods are based on diffusion generated by pressure gradients through a high speed flow of gases along a streamlined curvature. In a way these processes can be considered as non-rotating centrifuge processes. Enhancement of these forces is achieved by dilution of UF_6 with a carrier gas, hydrogen. This results in a higher flow velocity than that achieved with pure UF_6. Two processes based on this principle were developed.

In the German process, called the nozzle process, developed by Becker, a mixture of UF_6 and hydrogen (or helium) is compressed and then directed along a curved wall at a high velocity. The high speed gas experiences in the curved nozzle a force equal to 160 million times the gravitational force. As a result the heavier U-238 bearing molecules move preferentially to the wall relative to those containing U-235. At the end of the deflection, the gas jet is split by a knife-edge into the light and heavy fractions, which are withdrawn separately. A tube with about 80 such stages can achieve a separation factor of 1.0148. About 740 such stages would be required to produce 3% enriched product. A demonstration plant was set up in Brazil as part of the country's nuclear energy programme. The power requirement was 3,000 kWh/kg SWU.

The second process, called the Helikon Process or the Vortex Process, was developed by South Africa. In this process, a mixture of UF_6 and hydrogen is injected tangentially into a tube, which tapers to a small exit aperture at one or both ends. As the gas travels down the tube the gas undergoes separation under the centrifugal force, with the U-235 component concentrating near the axis and is withdrawn from the central part of the other end of the tube. Separation is caused by the centrifugal force. A 3% enrichment is achieved with 100 modules. The power energy requirement was 3,000–3,500 kWh/kg SWU.

In view of the high power requirements neither of the processes is in use.

6.4.8 Gas Centrifuge Method

Though the use of centrifugal fields for isotope separation was suggested as early as 1919 by Lindeman and Aston, it was not until 1934 that the technique was successfully employed by J.W. Beams for the separation of chlorine isotopes. The method, further explored in the Manhattan Project through the construction of a pilot project, was abandoned because "of the magnitude of the engineering problems involved." The process is now highly developed and is currently, the most preferred technology to produce both LEU and HEU. Its unique features include relatively low-energy consumption, short equilibrium time, and modular design.

In the gas centrifuge process, gaseous UF_6 is fed into a cylindrical rotor (1.5–5m long and 0.20m dia.) inside an evacuated casing, spinning at high speeds with the outer wall of the spinning cylinder moving at 400–500m/sec to give million times the acceleration of gravity. This drives the gas to occupy only a thin layer close to the rotor wall, and move with approximately the speed of the wall. The centrifugal force causes the heavier $^{238}UF_6$ molecules tend to move closer to the wall than the lighter $^{235}UF_6$ molecules, thus, partially separating the uranium isotopes. This separation is magnified by a relatively slow axial countercurrent flow of the gas within the centrifuge, which concentrates the lighter isotope enriched gas at the top end and the depleted gas at the other. The enriched product is withdrawn from the top of the rotor and the depleted or the waste stream is withdrawn from the bottom end through scoops that open opposite to the direction of rotation (Fig.6).

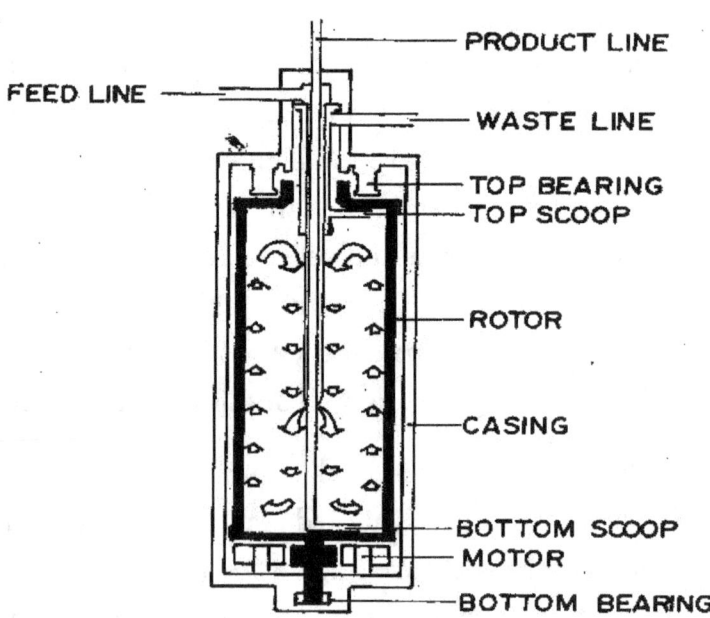

Fig. 6: Gas Centrifuge for Uranium enrichment. (Schematic)

Since the separation factors obtainable from a centrifuge range from 1.026 to 1.233 depending on the rotor, the number of stages for enrichment by centrifugation may be as few as 10 to 12, instead of a thousand times more for diffusion for achieving the given degree of enrichment. These stages are arranged in a cascade similar to those for diffusion. The enriched gas from one stage forms part of the feed for the next stage while the depleted UF_6 goes back to the previous stage. But compared to the gaseous diffusion, the capacity of a single centrifuge is much smaller. A single centrifuge might produce about 30g 0f HEU per year. For this reason, several more cascade units are required to achieve the required quantity of the enriched material. For commercial level separation the centrifuge stages normally consist of a large number of centrifuges connected in series as well as in parallel. (Fig.7). A cascade of 850–1,000 centrifuges 1.5m long operating continuously at 400m/sec would be able to produce about 20–25 kg of HEU in a year, enough for one weapon. A major advantage is the economy in space and electrical power. A facility capable of producing one bomb per year requires only about 600 square metres of floor space. The power requirement is only 50–60KWh per SWU compared to 2,400 KWh/SWU for gaseous diffusion. A modern centrifuge runs for more than 10 years with no maintenance.

India has been pursuing the gas centrifuge enrichment programme since the 1970s under the code name 'Rare Materials Project.' A small unit has been operating since 1885 at BARC, Trombay. A bigger facility located at Rattenhalli, 19km from Mysore in Karnataka has been functioning since 1990. The number of centrifuges in operation in 2006 was estimated between 2000–3000. The HEU is reportedly intended for fabricating the fuel for a nuclear submarine.

Fig. 7: Gas Centrifuge Cascade

6.4.9 Laser Separation Method

Laser separation methods involve the use of suitable lasers to selectively excite atoms or molecules containing one isotope of the atom or molecule, followed by separation. Current techniques for uranium enrichment fall into two categories. The first process, where the medium is uranium vapour, is called Atomic Vapour Laser Isotope Separation (AVLIS). In the second process, the medium is the vapour of uranium hexafluoride and it is called Molecular Laser Isotope Separation (MLIS).

Atomic Vapour Laser Isotope Separation (AVLIS) is based on the fact that U-235 and U-238 atoms absorb light of slightly different frequencies. The absorption wavelengths of these isotopes differ only by a minute amount (4.2×10^{-5} eV = 0.00295nm = 0.0295A^0). The dye lasers used in AVLIS are tuned so that only U-235 atoms absorb the laser light. Uranium vapour is generated by heating uranium metal with an electron beam. As molten and vapourised uranium are extremely reactive, the metal ingot is kept in a water-cooled crucible. The vapour stream is then laser-illuminated by radiation, which is selectively absorbed by U-235. Three wavelengths of a red-orange light of dye lasers are used as a radiation source for selectively ionising the U-235 atoms, leaving the other isotopes unaffected. The U-235 ions are then collected on negatively-charged surface units inside the separator unit. The product material condenses as a liquid, which then flows to a caster where it solidifies as a nugget. The tailings, which pass through the product collector, condense on the tailings collector and are removed.

An experimental AVLIS separation plant with an output of around 10^3 kg of LEU per year was built in 1990s, at the Lawrence Livermore National Laboratory, California, US. The technology was later transferred to US Enrichment Corporation (USEC). Avco Everett Research Laboratory and Exxon Nuclear Co., also engaged in the activity, claimed a high separation factor of 140 in their experiments. In this process enrichment to 3% U-235 may require only a single stage. Power consumption also is estimated to be low at 100 kWh per SWU. The process, however, requires very sophisticated hardware constructed from specialised materials for reliable operation in a harsh environment over extended periods of time. No country has adopted this technique so far for large-scale enrichment though several countries – US, France, Japan, Brazil, India, and Iran to mention a few, are believed to be working on this technology.

The MLIS process consists of two steps. In the first step, UF_6 is irradiated by an infrared laser system operating near 16μm wavelength, which selectively excites the $^{235}UF_6$, leaving the $^{238}UF_6$ relatively unexcited. In the second step, photons from a second laser system (IR or UV) preferentially dissociate the excited $^{235}UF_6$ freeing one fluorine atom to form $^{235}UF_5$. The $^{235}UF_5$ formed from dissociation separates from the gas as a powder, which can be filtered from the gas stream. MLIS is a stage-wise process, each stage requiring conversion of the enriched UF_5 product back to UF_6 for further enrichment. In view of the complexities involved the process, it does not appear to have made much headway.

The SILEX process, an acronym for Separation of Isotopes by Laser Excitation, has been developed by Australian scientists. While complete information on it is not available, the process uses a mixture of UF_6 and a carrier gas. The gas mixture is cooled to separate the resonance peaks for the isotopes. An infra-red laser of one or more frequencies selectively excites and ionises the U-235 isotope and not the other. Passing the mixture through an electric field will then separate the isotopes into a product stream and a tailing stream. GE Energy (US) entered into a partnership with Hitachi and CameCo in 2006 to commercially develop the SILEX process and is currently running a facility to evaluate the commercialisation of the process. A pilot plant was expected to become operational by 2012. If these efforts bear fruit an annual capacity of 3.5 to 6 million SWU is expected to be achieved during the coming years. The SILEX process is a low cost energy-efficient scheme, which may provide significant commercial advantage over the other competing laser processes.

The enrichment of U-235 from natural levels of 0–72% to 3–5% is achieved in a few stages with the laser technique, representing a great improvement over thousands of stages required for other methods (e.g., gas diffusion and centrifugation). Laser enrichment technology could become a potential alternative to gaseous diffusion and gas centrifuge techniques in the future. A commercial laser plant will use a little less than 100 enrichment units compared to more than 150,000 centrifuge machines. This translates into a much smaller plant as well as low production costs. Also, the laser process requires about 30% less natural uranium to produce a comparable amount of enriched product. Laser enrichment technology is even claimed as "the future of uranium enrichment."

6.4.10 Other Methods

Enrichment by chemical exchange developed by France is referred to as CHEMEX process. It is based on the exchange reaction that takes place between the trivalent and the tetravalent states of uranium ions in aqueous solution. Isotopic enrichment results from the tendency of U-238 to concentrate in the trivalent compound while U-235 concentrates in the tetravalent compound. The enriched form is removed by separating the tetravalent form using a suitable solvent such as TBP diluted with an aromatic solvent. A countercurrent extraction method is employed for this purpose.

Enrichment by ion-exchange was developed by the Ashahi Chemical Co. Japan by exploiting the chemical isotope exchange effect between the tetravalent (U^{+4}) and hexavalent (UO_2^{++}) states of uranium. The ion-exchanger is a spherical bead of porous ion-exchange resin with very high separation efficiency and an exchange rate over 1,000 times faster than the rates in most commercially available exchange resins. The aqueous phase flows through the stationary resin held in a column. The exchange between the un-absorbed uranium flowing through the band and that adsorbed on the resin, enhances the isotopic separation. In this continuous separation system, U-235 and U-238 tend to accumulate respectively at the entrance and exit ends of the adsorption column.

Though the above methods are under study in various countries, there are still no reports of their commercial scale exploitation for enrichment.

6.4.11 Commercial Level Enrichment

Large commercial enrichment plants are in operation in France, Germany, Netherlands, UK, US, and Russia, but four conglomerates control nearly all of world's LEU production. They are Tenex (Russia),

Areva (France), Urenco Group (Germany, Netherlands and UK) and USEC (US). The world enrichment capacity is detailed in Table 1. The current trend in enrichment technology is to retire obsolete diffusion plants. New centrifuge plants are also being built in France, US and elsewhere adding to the capacity. The contribution (per cent) of enriched uranium from gaseous diffusion, gas centrifuge and laser technique in 2007 was 25, 65 and nil respectively. By 2017, the respective contributions are projected to be nil, 93 and 3. Enriched uranium recovered from dismantled nuclear weapons under international agreements is currently an important source of LEU. In 1993, the US and Russia have entered into an agreement by which Russia will convert, over a period of 20 years, 500 metric tonnes of HEU from the dismantled nuclear weapons to LEU and supply to US commercial power reactors. According to a 2009 report, Russia has already converted 382 metric tonnes of HEU to 11,047 metric tonnes of LEU. Most of this will go into the operating reactors and the upcoming reactors in the US.

Table 1: World Enrichment Capacity (thousand SWU/year) (Operational and planned)

Country	2010	2015	2020
France (Areva)	8,500#	7,000	7,500
UK (Capenhurst)			
Netherlands (Almelo)			
Germany (Urenco, Gronau)	12,800	12,800	12,300
Japan (JNFL)	150	750	1,500
US (USEC)	11,300#	3,800	3,800
US (Urenco)	200	5,800	5,900
US (Areva)	0	0	3,300
US (Global enrichment)	0	2,000	3,500
Russia (Tenex)	23,000	33,000	33–35,000
China (CNNC)	1,300	3,300	6–8,000
Pakistan, Brazil, Iran, India	100	300	300
TOTAL (app)	57,350	68,000	74–81,000
REQUIREMENTS	48,890	56,000	66,535
# – diffusion			

Source: WNA Market Report 2009; WNA Fuel Cycle Enrichment plenary session WNFC April 2011

6.5 Depleted Uranium (DU)

The residual part of enrichment, called depleted uranium (DU) contains about 99.8% U-238 and 0.2% U-235 (by mass). Producing one kg of 5% enriched uranium requires 11.8 kg natural uranium, leaving about 10.8 kg of DU. The total estimated DU stocks lying with various countries (US, UK, Russia, France, Japan, China, S. Korea and S. Africa) in 2002 are placed at 1.2 million metric tonnes. Depleted uranium is stored over long periods as UF_6 or preferably by reconversion to U_3O_8. By early

2007, one quarter of the DU has been reconverted. The ownership of the DU is normally retained by the enrichment firm as part of the commercial deal. The DU is a long-term strategic fertile material for use in the future generation of fast breeder reactors. The DU will also be a usable resource when technology is available for more efficient enrichment or when the uranium prices rise, say, above $500 per kg. The enrichment companies with large quantities DU are thus owners of a potential bonanza.

Meanwhile DU, with its high density (1.7 times lead), is finding extensive use in armour-piercing projectiles. The projectiles are made with an alloy of DU and small amounts of a metal such as molybdenum or titanium to give extra strength and resistance to corrosion. Because of their long range and high penetrating power these projectiles are called kinetic energy penetrators. These projectiles have been extensively used in the wars in the Gulf (290,000 kg of DU), Balkans (12,700 kg of DU) and Iraq (about 140,000 kg of DU). Much of the material used would have fractured and disintegrated on impact, dispersing as long-lived radioactive uranium particles into the air and soil, putting the soldiers and civilian population to risk.

At least 17 countries are believed to have DU weapons in their arsenals. These include UK, US, France, Russia, Greece, Turkey, Israel, Saudi Arabia, Bahrain, Oman, Egypt, Kuwait, Pakistan, India, Thailand and China.

6.6 Centrifuge and Laser Enrichment Technologies and Nuclear Weapons

Proliferation

To begin with, a few technologically-advanced rich nations had control over building nuclear weapons because the uranium enrichment process was enormously complicated and expensive. But with the development of gas centrifuge and laser techniques for uranium enrichment the risk of nuclear weapon proliferation is definitely on the rise. Gun-type nuclear weapons using enriched uranium (HEU) are the easiest of all nuclear weapons to design and build. The cheaper and less complicated gas centrifuge enrichment technology already has become a cause for international nuclear proliferation. That even the best efforts to safeguard sensitive technologies can and will eventually fail has been proved by the Pakistani scientist A.Q. Khan who clandestinely procured the enrichment technology from UCN, the Dutch partner in the Urenco uranium enrichment consortium and other components from various suppliers that went into the construction of the enrichment plant at Kahuta culminating with the acquisition of nuclear weapon capability by Pakistan. Khan did not stop with this. With the Pakistani Government's complicity, he extended assistance to North Korea, Iran, Libya, and some unknown countries. These activities have been well chronicled by various agencies (e.g. US CRS Reports). Currently, Iran is actively engaged in enrichment activity at Nataz, flouting the NPT of which it is a signatory. Iran has announced its aim to install 50,000 centrifuges for this purpose and even announced in February 2010 its intention to enrich uranium to 20% level, up from 3.5% required for a nuclear power plant. From 20% enrichment, Iran could make a quick leap to enrich this uranium to the weapons grade.

Experts say that the uranium enrichment by the SILEX laser technology could be several times more efficient than the centrifuge technique. If this technology, which is less elaborate falls into the hands of rogue nations, it could give greater impetus to nuclear weapons proliferation.

6.7 The Hexapartite Safeguards Project (HSP)

With the availability of less complex gas centrifuge technology, uranium enrichment plants have become one of the greatest challenges to the nuclear non-proliferation regime. Through building a large enough gas centrifuge enrichment plant (GCEP), a state could amass the necessary material for one nuclear weapon in a short span of time as was demonstrated by Pakistan. With the object of evolving effective safeguards for GCEPs, six countries having gas centrifuge technology (US, UK, Germany, Netherlands, Japan and Australia) along with the International Atomic Energy Agency (IAEA) and the European Atomic Energy Community (EURATOM) reached a consensus that enriched uranium production (LEU and HEU) was a great proliferation risk. To control the proliferation they focused on verification techniques on the detection of HEU and evolved in 1983 the Hexapartite Safeguards Project (HSP). The Project adopted several measures which only the HSP participants were initially committed to implementing. These measures were later used as a model for the safeguards applied to additional states. Under this programme, the IAEA continuously investigates new measures and incorporates new technologies to meet the HSP goal of detecting diversion or misuse of declared material at declared facilities. The updated IAEA model safeguards for gas centrifuge plants are:

- Timely detection of the diversion of significant quantities of natural, depleted, or low enriched UF_6 from the declared flow through the plant and the deterrence of such diversion through early detection.

- Timely detection of the misuse of the facility for producing undeclared product from undeclared feed and the deterrence of such misuse through early detection.

- Timely detection of the misuse of the facility to produce UF_6 at enrichment higher than the declared maximum, in particular HEU, and the deterrence of such misuse through early detection.

As a part of the objective of preventing nuclear proliferation, the International atomic Energy Agency (IAEA) approved in December 2010, a proposal for the establishment of International Uranium Enrichment Centres (IUERC) which would act as low enriched uranium (LEU) banks to be run by the Agency. Under this proposal, the IAEA will be the owner of LEU in the bank, which will be located in one or more member states that are prepared to act as host states. The LEU from the bank will be supplied to member states that experienced a nuclear fuel disruption due to exceptional circumstances and are unable to secure supply from the commercial market. The IEUC charter states that to avail this facility the participant states must forego development of domestic enrichment capabilities. The first centre is located near the Siberian city of Angarsk in Russia.

Presently, the IAEA safeguards are under greater stress than at any other time. Compliance concerns, a shortage of resources and technology and growing responsibilities threaten to undermine the effectiveness and credibility of this vital and fundamental pillar of the non-proliferation agenda.

7

Nuclear Power Reactors

7.1 Introduction

December 2, 1942, is a historic date in the nuclear era. As described earlier, on this day Enrico Fermi demonstrated the feasibility of a sustained and controlled nuclear fission chain reaction in a nuclear pile assembled at the University of Chicago. This demonstration opened up two possibilities: building nuclear weapons of unprecedented destructive power and exploitation of nuclear energy for civilian use.

With the world war raging at the time, building nuclear weapons naturally received priority. However, with the cessation of hostilities, programmes were initiated for the development of nuclear technology for power generation. The US had a technological headstart in this field thanks to the wartime Manhattan Project and took the lead in this area too. Western Europe, the cradle of nuclear research in the pre-war period, also wanted to regain its lost ground in this area. Six countries including the US, UK, France, USSR, Canada, and Sweden pioneered the development of nuclear technology for power generation. In due course, other countries received assistance from these countries in the installation of nuclear power plants to meet their power requirements.

Rising from less than 1 GWe in 1960, nuclear power production rose to 100 GWe in the late 1970s and then to 300 GWe in the late 1980s. Since the late 1980s, the worldwide production rate rose much more slowly, to reach 366 GWe in 2010 from 439 nuclear reactors in 31 countries, providing about 16% of the world's electricity. Worldwide nuclear electrical power production in 2009 is put at 2558 TWh. The US produces the most, with nuclear power providing 20.2% of its electricity in 2009. France produces the highest percentage (75.2%) of its power needs. The European Union, as a whole, gets 30% of electricity from nuclear energy. As of December 2012, over 15,000 reactor-years of experience in civilian nuclear power production has been gained worldwide.

Nuclear fission has several significant features that make nuclear power production very attractive. Unlike fossil fuels, it does not require oxygen nor does it produce carbon dioxide (greenhouse gas). It produces extraordinarily high energy density per unit weight of the fuel. Just 1kg of uranium-235, if completely fissioned, would give the same thermal energy (heat) as 2,500,000 kg of coal. Nuclear power plants will be economical in places where the transport of conventional fossil fuels would be expensive. Nuclear powered submarines can cruise for a long duration without the need for refueling.

7.2 Anatomy of a Nuclear Power Reactor

A nuclear reactor is a device for controlled energy production through fission of the nuclei of certain heavy elements (uranium-235 and synthetic plutonium-239). In a nuclear power reactor, the thermal energy released is used to make steam which drives a turbine to generate electricity. The conversion factor for thermal energy into electrical energy is generally 35 to 45%. The thermal electrical power output of a reactor is expressed as MWt and the electrical power output as MWe.

The civilian power reactor technology has evolved from the early (1950–1960) research reactors to its present state of commercial reactors. The principles involved in construction are the same for most types of reactors. The principal components of a reactor are:

- Fuel assembly
- Moderator
- Coolant
- Control device
- Supporting structure
- Pressure vessel – horizontal or vertical
- Steam generator

7.2.1 Fuel Assembly

Currently, uranium is the most commonly used nuclear fuel in reactors. Naturally occurring uranium contains two major isotopes; – U-235 (0.72%), U-238 (99.28%) and U-234 (0.0054%). Of these, U-235 is mainly responsible for the fission reaction. Uranium fuel is usually in the form of ceramic uranium oxide, UO_2, (M.P. 28000 C). The fuel pellets (1 cm across and 1.5 cm long) are contained in a long tube (3.5 meters long) made of a corrosion resistant and neutron permeable alloy like Zircaloy to form a fuel rod or a pin. The reactor core is a fuel assembly consisting of the required number of bundles, each containing 50–200 pins. The number of fuel assemblies in a reactor depends on its capacity.

Uranium-235 undergoes fission with a thermal neutron releasing energy. The fission chain is sustained through the controlled release of more than one neutron in each fission event. The U-238 in the fuel captures a thermal neutron and forms the heavier isotope, U-239, which through the emission of an electron is converted to neptunium-239 (^{239}Np). Emitting an Np-239 is converted to plutonium-239 (^{239}Pu) which is fissionable. For this reason, U-238 is known as a fertile material. Part of the plutonium formed in the fuel undergoes fission contributing to the energy generation.

$$_{92}U^{238} + {_0}n^1 \rightarrow {_{92}}U^{239} \rightarrow {_{93}}Np^{239} + {_{-1}}e^0; \; {_{93}}Np^{239} \rightarrow {_{94}}Pu^{239} + {_{-1}}e^0$$

Natural uranium is used as the fuel with graphite or heavy water as a moderator. When ordinary water is the moderator, low enriched uranium (LEU) is used as the fuel to make up the neutron loss through absorption by ordinary hydrogen. A majority of the present-day reactors run on low-enriched (1.8 to 3.5%) uranium. The enrichment level of uranium fuel, the geometry of the bundles and the type of loading – vertical or horizontal – depend on the type of reactor. Plutonium is also used as fuel in some reactors.

When a new reactor is being brought into operation, a neutron source is used to initiate the nuclear chain reaction. An alpha emitting element such as radium or polonium mixed with beryllium is used for this purpose. The neutrons released through (α,n) reaction initiate the fission chain reaction. Restarting a reactor does not normally require a neutron source as there will be sufficient neutrons available in the reactor.

Uranium-233, another man-made isotope also exhibits fission. This isotope is formed from thorium-232 by irradiation with neutrons. It does not have any use to a significant extent at present, but maybe a fuel of choice in the future.

7.2.2 Moderator

The energy released in nuclear fission is carried by the fission fragments and the neutrons generated. The fission fragments being relatively heavy are trapped within the fuel pellets while the neutrons pass through the Zircaloy cladding. The neutrons in the nascent state have a high kinetic energy, travelling with a speed of about 20,000km per second. These neutrons are called 'fast' neutrons. For sustaining the chain reaction in the fissile material their speed has to be reduced to about 2.5km per second. The neutrons thus slowed down, are called 'thermal' neutrons. For achieving the thermalisation of the neutrons, the fuel elements in a reactor are surrounded by a material called a moderator. A light element or a compound containing a light element acts as an efficient moderator. The fast neutrons lose their energy through repeated elastic collisions with the nuclei of light elements in the moderator and reach thermal energy levels. Only nuclei of light elements (hydrogen to carbon) which do not significantly absorb neutrons serve as efficient moderators.

The commonly used moderators are ordinary water (H_2O), heavy water (D_2O, D = Deuterium – the heavier isotope of hydrogen with mass 2) and graphite (carbon). Heavy water occurs in natural water at a concentration of 150 molecules to one million water molecules. The first two, being liquids, are easy to handle. Apart from moderating the neutrons, hydrogen in ordinary water also absorbs slow neutrons to some extent, causing some neutron loss. However, in view of its easy availability and low cost it is the preferred moderator in over 70% of the reactors in operation. When light water is used as a moderator, light enriched uranium (LEU) is used as fuel to compensate for the neutron loss. These reactors are called Light Water Reactors (LWR).

Heavy water has a lower tendency to absorb neutrons. For this reason it acts as a better moderator and can be used in reactors using natural uranium as fuel. Such reactors are known as Heavy Water Reactors

(HWR) or CANDU (proprietary name) reactors. Pure graphite, which does not absorb neutrons, lies between water and heavy water in its moderating capability. It was used as a moderator in early reactor designs (e.g. Magnox reactors).

The effective performance of a moderator is given in terms of its moderating ratio, which is defined as ratio of the macroscopic neutron slowing down power to the macroscopic cross section for neutron absorption. The higher the moderating ratio the more effectively the material performs as a moderator. The moderating ratios of the commonly used moderators are in the order: light water – 72, graphite – 188 and heavy water – 21,000.

While countries operating light water reactors need facilities for isotopic enrichment of uranium or import of enriched fuel, those using heavy water have to have infrastructure for production of heavy water in sufficient quantity at a reasonable price or depend upon import.

Reactors based on fast neutron fission of plutonium do not need a moderator. Such reactors are called 'fast neutron reactors.'

7.2.3 Coolant

The heat generated in nuclear fission in a power reactor has to be extracted for operating a turbine that generates electricity. The coolant which is a fluid plays a crucial role in this function. Depending on the reactor type, this can be water, heavy water or a gas like carbon dioxide or helium. When water or heavy water is used as coolant it plays the dual role of a moderator and a coolant. A gas (helium or carbon dioxide) is used as a coolant when a solid moderator like graphite is in place. Liquid metals (sodium or sodium-potassium alloy) with low moderating ratio are used as coolants in fast neutron reactors.

The coolant is heated by the nuclear reactions occurring inside the core. Water or heavy water, used as a coolant, is maintained at a very high pressure (1000–2000 psi, 7–15 MPa). In boiling water reactors (BWR) the water heated to boiling (285°C) under high pressure in the core gives out steam to drive the turbines for producing electricity. In pressurised water reactors (PWR), the pressure is high enough to keep water in the liquid phase even at a high temperature. A heat exchanger uses the super-heated coolant to generate steam. This steam, in turn, drives turbines to produce electricity.

The coolant also keeps the temperature of the fuel pellets and the cladding within limits and prevents their melting. The coolant and the metal cladding on the fuel pellets should be compatible so that they do not chemically react at the temperature prevailing outside the fuel pins.

7.2.4 Void Coefficient

Void coefficient or void coefficient of reactivity is a value that can be used to estimate how much the reactor activity changes as voids (or steam bubbles) form in the reactor moderator or coolant. If the reactivity is positive the core power tends to increase, if it is negative the core power tends to decrease and if the reactivity is zero the core power tends to remain static. Boiling water reactors have generally negative void coefficient and in normal operation this allows the reactor power to be adjusted by changing the rate of the moderator water flow through the core. Pressurised water reactors operate with no voids at all as the water serves as both moderator and coolant.

The large negative void coefficient ensures that if the heavy water boils or is lost, the power output drops. If the temperature inside the reactor rises the liquid coolant may boil causing voids inside the reactor, leading to a positive void coefficient. The temperature may also rise if the coolant is lost from the reactor due to an accident. If the void coefficient is large enough and the control systems do not respond quickly, this can form a feedback loop which can boil all the coolant in the reactor leading to the burn up of the reactor core. Such a rise in the positive void coefficient has led to the Chernobyl reactor accident. This RBMK reactor is moderated by graphite but cooled by water that flows over the fuel rods. The coolant water also acts as a neutron absorber. When the temperature rose, the water core boiled and there was a reduction in the absorption of neutrons by water vapour. This resulted in a rise in the availability of neutrons (moderated by graphite) causing more fissions. In addition, the graphite expanded due to a rise in temperature and this effectively increased its ability to thermalise the neutrons causing more fissions, generating more heat. Such a condition led to the burn up of the reactor core.

7.2.5 Reactor Control Device

A nuclear reactor is provided with many in-built controls for its smooth and accident-free operation. In an emergency such as a mechanical or structural failure, the fission reaction in the core has to be stopped quickly to prevent any major disaster. A periodic shutdown is also necessary for loading fresh fuel. In a running reactor, the rate of fission in the core has to be continually monitored and controlled so that it does not become unduly fast liberating too much heat, which could result in a fuel element meltdown. The control of a reactor is achieved through the use of control rods of high neutron-absorbing materials such as cadmium, or boron. Cadmium is used in the metallic form or as an alloy with silver and indium and boron as carbide. Other neutron absorbing elements hafnium, gadolinium, dysprosium, or their alloys also find use as control elements.

The control device is operated either manually or through automation. The control rods stand vertically within the core. When the rods are fully inserted into the core, all free neutrons are absorbed and the fission process stops. The number of control rods inserted and the degree of insertion regulates the reactivity level of the core. Provision is also made for the rods to drop automatically into the core in the event of an emergency. For water-cooled reactors, an additional provision is made for a solution of a neutron absorbing substance to flow automatically into the coolant channels. In the case of gas-cooled reactors, high-pressure nitrogen is injected into the primary coolant cycle to absorb the neutrons and stop the fission chain.

7.2.6 Container Vessel

A nuclear reactor is provided with many in-built controls for its smooth and accident-free operation. In an emergency such as a mechanical or structural failure, the fission reaction in the core has to be stopped quickly to prevent any major disaster. A periodic shutdown is also necessary for loading fresh fuel. In a running reactor, the rate of fission in the core has to be continually monitored and controlled so that it does not become unduly fast liberating too much heat, which could result in a fuel element meltdown. The control of a reactor is achieved through the use of control rods of high neutron-absorbing materials such as cadmium, or boron. Cadmium is used in the metallic form or as an alloy with silver and indium and boron as carbide. Other neutron absorbing elements hafnium, gadolinium, dysprosium, or their alloys also find use as control elements.

The control device is operated either manually or through automation. The control rods stand vertically within the core. When the rods are fully inserted into the core, all free neutrons are absorbed and the fission process stops. The number of control rods inserted and the degree of insertion regulates the reactivity level of the core. Provision is also made for the rods to drop automatically into the core in the event of an emergency. For water-cooled reactors, an additional provision is made for a solution of a neutron absorbing substance to flow automatically into the coolant channels. In the case of gas-cooled reactors, high-pressure nitrogen is injected into the primary coolant cycle to absorb the neutrons and stop the fission chain.

7.3 Evolution of Nuclear Reactor Systems

Advances in nuclear power reactor design technology aimed at maximising efficiency and safety can be sequentially seen as comprising four distinct generations of nuclear power reactors.

Generation I: These reactors are early prototype reactors.

Generation II: These are commercial nuclear power reactors currently in operation

Generation III: These are light water reactors and other systems with inherent safety features designed in recent years.

Generation IV: The next generation reactor systems to be designed and built during the coming decades.

7.4 Generation I Power Reactors

The first electricity-producing US prototype reactor near Arco, Idaho, went into operation in 1951 generating a power of 250 kWe, just sufficient to supply power to the reactor building. This project finally led to the launching of the first nuclear submarine Nautilus in 1955. A nuclear submarine can remain submerged for months at a time while a diesel submarine must periodically resurface or snorkel to recharge its batteries. About this time the USSR commissioned its nuclear power plant at Obninsk with a generating capacity of 5 MWe. This was the precursor of the later RBMK reactors which came into commercial operation after 1964. The first US nuclear power plant at Shippingport, Pennsylvania, a prototype light water fast breeder reactor for commercial power generation with a generating power of 60 MWe, began operations on December 2, 1957. This reactor was shut down in 1982. In 1960, the General Electric started the first commercial boiling water nuclear power plant at Dresden near Chicago, US. The Fermi I (a breeder prototype) reactor generating 94 MWe built in 1966 was shut down in 1972 after a meltdown.

The MAGNOX reactors are pressurised, carbon dioxide cooled, graphite moderated reactors using natural uranium as fuel with Magnox (MAGnesium Non OXdising) alloy as the cladding. Also called gas-cooled reactors (GCR), they were built during the period 1956–1971. The fuel rods are loaded into vertical channels in a core constructed from graphite blocks (Fig.1). The hot gas produces steam from water in a steam generator, located outside the reactor. The steam runs a turbine, which is a part of the electric generator. The entire reactor assembly is housed in a thick concrete containment structure, which acts as a radiation shield too.

The early MAGNOX reactors built in Calder Hall had a capacity of only 50MWe. They were primarily used to produce weapons-grade plutonium, though some power was fed into the grid. From 1964, all the reactors were mainly used for commercial power generation. Most of them are of 200MWe capacity, except the two units at Anglesey (490MWe). Two reactors of this type were sold to Japan and Italy. MAGNOX reactors are not very efficient. In the older design, the boilers and gas ducting are outside the biological shielding and this led to the exposure of operating personnel to relatively high gamma and neutron doses.

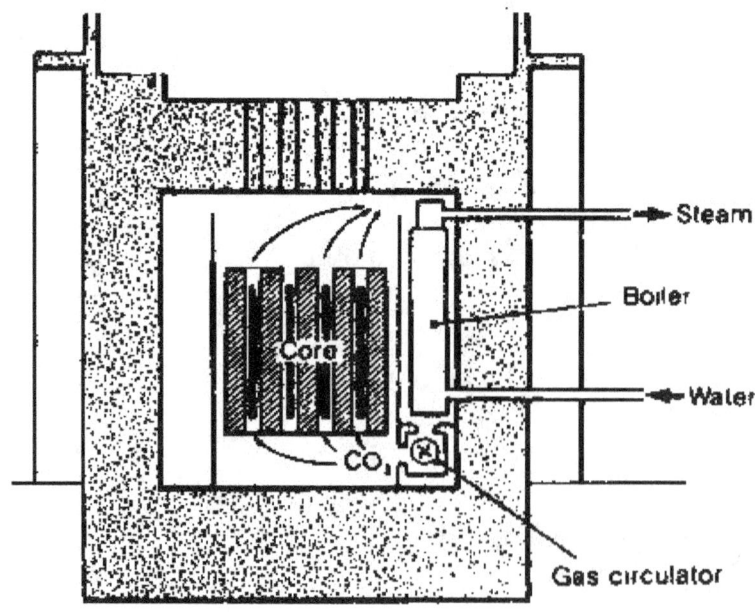

Fig. 1: MAGNOX Reactor

In all 11 power stations totaling 26 MAGNOX reactor units were built in the UK. All these units except the unit at Wylfa in Wales have been shut down. The Wylfa unit is scheduled to be shut down in 2014.

In 1962, the first Canadian nuclear power plant using uranium and heavy water became functional. The first nuclear power plant in France was built in 1965. Sweden built its first heavy water reactor in 1964.

7.5 Generation II Reactors

With the experience gained with Generation I reactors, technology was developed for the commercial power reactors of subsequent generations. Generation II nuclear reactors began operation in the 1970s. These comprise 400-odd commercial reactors, including the outmoded but still operating graphite moderated RBMK reactors.

7.5.1 Graphite Moderated Reactors

RBMK Reactor: About the same time as the MAGNOX reactors were being commissioned, the USSR developed the RBMK (acronym for Reactor Bolshoi Moschnost Kanalynyi) reactor design and constructed many commercial reactors of this type. The design consists of a large graphite core containing some 1700 vertical channels, each containing 1.8% enriched uranium dioxide. Heat is removed from the fuel by pumping up water under pressure through pressure tubes. This heated water goes to turbines via steel drums. As of today, there are at least 11 RBMK reactor units operating in Russia, two in Ukraine and two in Lithuania, with a total power output of 12.3 GWe. Most of the operating reactors have a generating capacity of 1,000 MWe each while some produce 1,500 MWe. The reactors suffer from a number of design and safety flaws, though modifications have been made to these reactors, there is international pressure to close down these reactors due to design shortcomings. The most significant design deficiency is the lack of a containment structure as the final barrier to isolate the dangerous radioactive material in the event of a serious accident (e.g. Chernobyl accident).

Advanced Gas-Cooled Reactor (AGR): The Advanced Gas-cooled Reactor (AGR) is a successor to the British MAGNOX reactor. As in a MAGNOX reactor, the moderator and coolant are graphite and carbon-dioxide, respectively. The fuel, however, is LEU uranium oxide pellets, enriched to 2.5–3.5% and loaded in stainless steel tubes. The first commercial AGR came online in 1976. The oxide fuel and the cladding permit operation of the reactor at a temperature of 640°C, which is higher than in many other reactors, yielding a higher thermal efficiency (electricity generated/heat generated) of 41%. The control rods penetrate the moderator. The secondary shutdown system involves injecting nitrogen to the coolant. Radiation exposure risk to the personnel is minimised by housing the steam generators and gas circulators within a combined concrete pressure/radiation shield. Currently, 14 AGRs are functioning in the UK with an average capacity of 600MWe each, generating a total power of about 8,400MWe.

7.5.2 Water Moderated Reactors

Of the commercial reactors in operation worldwide, approximately 80% are light water moderated and light water cooled reactors (LWRs). There are mainly two types of LWRs. One is the Pressurized Water Reactor (PWR) in which water is the moderator cum coolant in the reactor core. It is kept under sufficiently high pressure to suppress its boiling. Steam is generated in a heat exchanger and sent to the turbine for electricity generation. The other is the Boiling Water Reactor (BWR) in which the coolant water in the reactor is allowed to boil and the resulting steam is directly sent to the turbine.

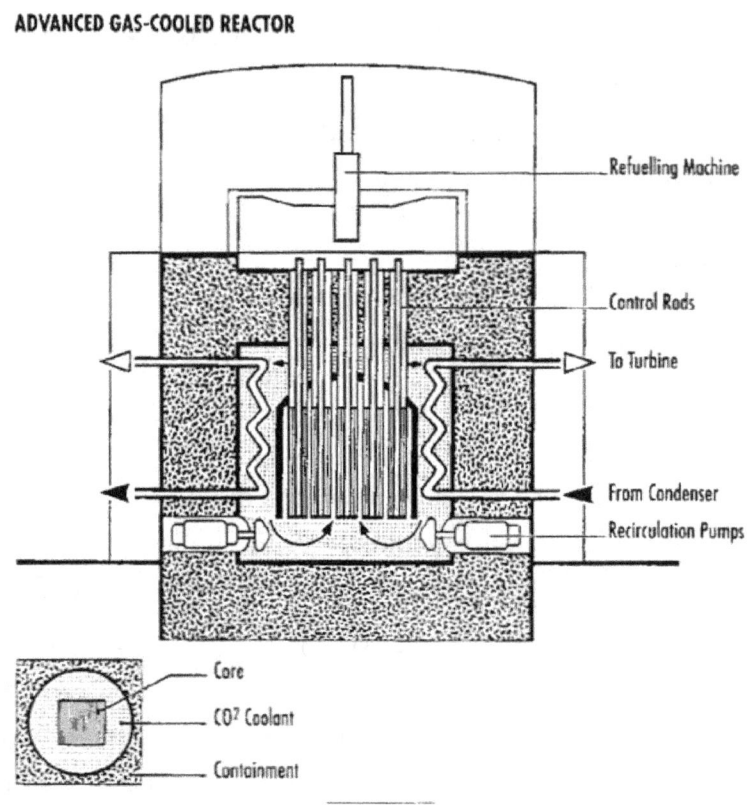

Fig. 2: Advanced Gas-cooled Reactor (AGR)

Pressurized Water Reactor (PWR): Currently the most widely used reactor type in the world is the Pressurized Water Reactor (PWR). This reactor was originally designed in the US by Westinghouse Bettis Atomic Power Laboratories for supplying power in submarines and aircraft carriers. Westinghouse Nuclear Power Division later adapted this model for commercial nuclear power plants. The first commercial plant was set up at Shippingport, Pittsburgh, US in 1982.

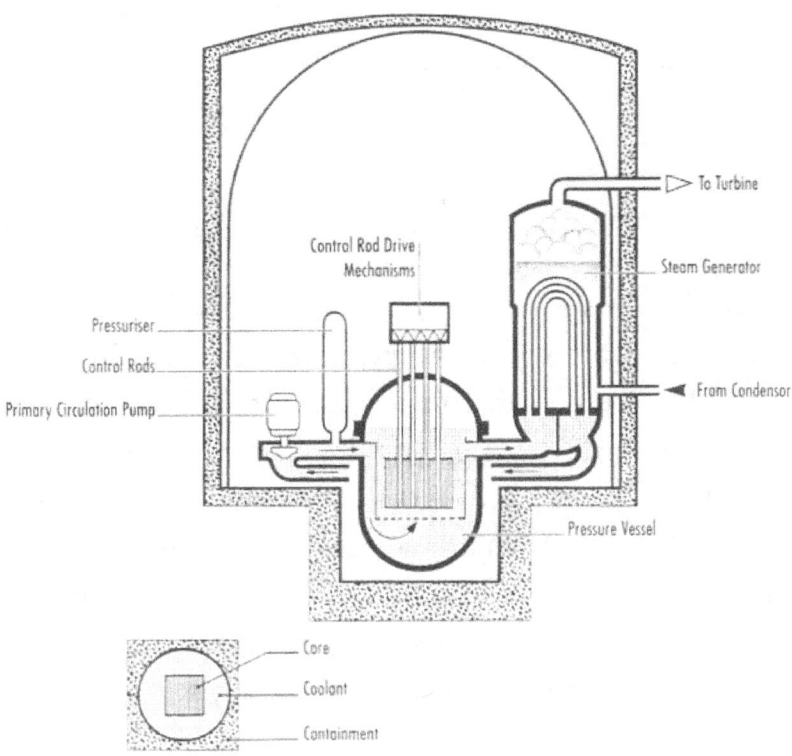

Fig. 3: Pressurised Water Reactor

The fuel in the PWR reactor (Fig.3) is low-enriched (about 3.2% U-235) uranium dioxide packed in the form of pellets in tubes made of Zircaloy. A large PWR produces 900–1,500MWe with 80–100 tonnes of uranium loading in 150–250 fuel assemblies containing 200–300 rods each. The entire fuel assembly is kept in a steel vessel, which in turn is located in a concrete containment building. The primary coolant, which also serves as a moderator, is pure water which is pumped through the reactor core under high pressure from the bottom. The inlet water is at 275°C and the outlet water at 315°C. The outlet water under high pressure (150 atmospheres) passes through a steam generator circuit, in which steam is generated to drive the turbine.

A significant drawback of the PWR is accelerated corrosion of the components due to water circulating under high temperature and pressure. For this reason, the steam generators, in particular, need frequent replacement.

As the steam generator of a PWR is housed in the same concrete containment structure as the core, the entire radioactive contamination from the primary coolant remains sequestered in the building without affecting the turbine, condenser, pumps, etc. For example, when the 900 MWe PWR unit at Three Mile Island was involved in an accident nearly all the radioactivity released was sequestered in the containment building with negligible release into the environment. Yet another advantage is the inherent negative temperature coefficient of reactivity (the reaction slowing down as the temperature rises unduly) of the system. Any increase in temperature inside the core due to malfunctioning of controls causes the water to expand and become less dense paring down neutron moderation and controlling the reactivity of the reactor.

PWRs from Russia (former the Soviet Union) are called VVERs (water-water-energy-reactor) with 440 MWe and 1000 MWe outputs. These reactors, which conform to international standards, are exported to several East European countries and also to India. Europe-based Areva/Framatome NP also manufactures PWR reactors.

Worldwide 265 commercial PWRs are in operation with an aggregate power output of 2,516 GWe.

Boiling Water Reactor: The Boiling Water Reactor (BWR) concept was developed in the mid-1950s by the General Electric Company, US in collaboration with several US National Laboratories. The design has many similarities to the PWR except that light water acts as both moderator and coolant in this reactor. The core consists of low-enriched (1.5–3%) uranium oxide pellets clad in long stainless steel tubes. The fuel rods, about 36–49, are bundled together and arranged in a vertical lattice. A modern large capacity BWR holds about 140 tonnes of uranium. Pre-heated high purity light water under pressure (75atmospheres) flows from the bottom into the core. The BWR does away with a steam generator and allows the nuclear heat to generate steam under the high temperature (285°C) and high-pressure conditions in the core. The primary steam is extracted by steam separators and fed directly to drive a turbine. Steam coming out of the turbine is cooled in a condenser and the hot water is returned to the reactor core completing the loop (Fig.4).

The reactor power is controlled in two ways: inserting/withdrawing the control rods or changing the water flow-rate through the core.

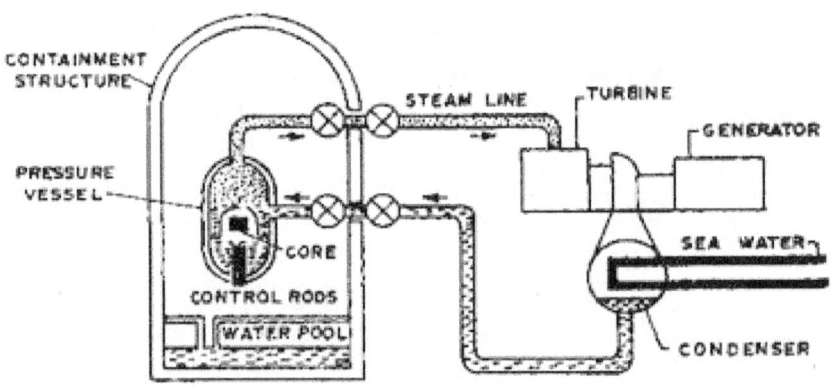

Fig. 4: Boiling Water Reactor

The water around the core invariably carries traces of radioactive nuclides. A part of the activity enters the steam circulator and the turbine house (through steam). Hence, these parts need shielding during normal operation. Radiological protection to personnel is required during maintenance. A containment structure, 1.2 to 2.4m thick steel reinforced, pre-stressed concrete that houses the reactor seals it off in an emergency.

When compared to the Pressurized Water Reactor (PWR) the design of a BWR is simpler with a higher thermal efficiency. But these advantages are offset by expensive shielding requirements and maintenance.

Apart from the General Electric Co., other companies that supply BWRs are Areva (France), Siemens (Germany) and Ŝkoda JS (Czech Republic).

Next to PWR the BWR is the most common reactor in use. A total of 94 reactors of BWR type with 86.4 GWe output are currently in commercial operation worldwide.

7.5.3 Pressurized Heavy Water Reactors (PHWR)

It is possible to use natural uranium as reactor fuel with, instead of light water, heavy water as a moderator. While the US, which had developed uranium enrichment facilities in the Manhattan Project, opted for light water-enriched uranium reactors in its post-World War plans, countries with no enrichment facilities at the time directed their efforts to the construction of natural uranium fuelled reactors with heavy water as a moderator. In the 1950s, The Atomic Energy of Canada commercialised the CANDU (CANada-Deuterium-Uranium) reactor, a pressurised heavy water reactor designed by a consortium of Canadian Government and private industry.

In contrast to PWR, the PHWR uses pressure tubes instead of a large pressure vessel to hold the fuel bundles. A horizontal vessel called Calandria houses heavy water with a grid of Zircaloy-2 pressure tubes. Fuel pins, containing natural uranium oxide pellets clad in Zircaloy-2 in the form of short bundles, are loaded in the pressure tubes through which the high temperature-high pressure heavy water flows to extract the heat. The coolant pressure is about the same as in PWR. The hot coolant is circulated through a bank of steam generators. The whole system is housed in a strong containment structure (building). As in the case of PWR, the primary heavy water system, which becomes radioactive over time, does not leave the reactor's containment building (Fig.5). The steam generator tubes are of a nickel alloy. A distinctive feature of CANDU/PHWR is the extensive use of carbon steel. The chemistry of both the primary heavy water coolant and the secondary light water has to be regulated to take care of this mixed metal system under severe operating conditions. On-line refueling by a remote method is one of the unique features of the CANDU system. The core is designed to be continuously 'stoked' with new fuel and not through a total replacement as in a batch process. Apart from increased reactor time availability, this reduces excess reactivity in the core and also the need for burnable poisons. A burnable poison is a neutron absorber in the fuel that controls excess fuel reactivity, especially in the initial stages. Incorporated in the fuel or fuel cladding of the reactor, it gradually burns up the poison products under neutron irradiation and converts them into materials of low neutron absorption. Boron or gadolinium compounds are used as burnable poisons for the purpose.

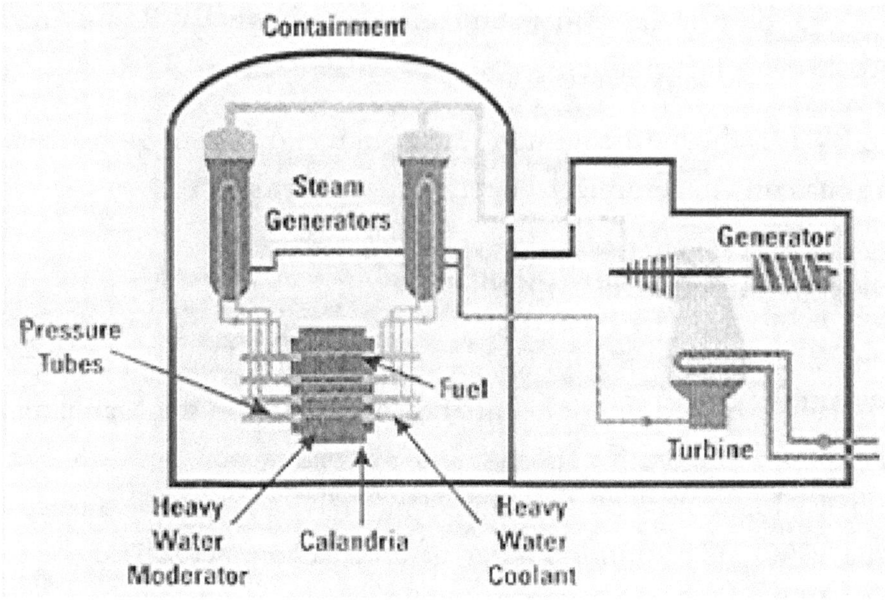

Fig. 5: PHWR Reactor

A major disadvantage of this reactor is the need for infrastructure for a viable large-scale production of heavy water or its import.

Apart from using exclusively for its domestic nuclear power supply, Canada exported the CANDU type reactors to India, China, Pakistan, Argentina, South Korea, and Romania. These 44 reactors have output capacities ranging from 200 MWe to 935 MWe. The total power output of the CANDU reactors worldwide is 24.3 GWe.

7.5.4 Fast Neutron Reactors (FNR)

The Fast Neutron Reactors (also called Fast Breeder Reactors, FBR) represent a technological advance over the conventional thermal reactors. Fast neutron reactor designs with sodium cooling have been under development from the very beginning of nuclear power production in the 1940s. They were even heralded as the solution to all energy problems, since theoretically, the breeder reactors permit the complete use of uranium as a fuel by systematically breeding fissile Pu-239 out of the non-fissile U-238, thus increasing uranium energy reserves a factor of about 100.

In a conventional thermal reactor, the fuel is either natural uranium or LEU. While the U-235 does most of the fissioning, more than 90% of the uranium atoms in the fuel are U-238. These absorb thermal neutrons and undergo conversion to plutonium-239 atoms.

$$_{92}U^{238} + _{0}n^{1} \rightarrow {}_{92}U^{239} \rightarrow {}_{93}Np^{239} + {}_{-1}e^{0}; \; _{93}Np^{239} \rightarrow {}_{94}Pu^{239} + {}_{-1}e^{0}$$

The product Pu-239 which is fissionable with neutrons can be separated and used as reactor fuel. But as the reactor operates, plutonium-239 also undergoes fission giving heat. The longer the fuel elements remain in the running reactor the greater is the proportion of power generated from plutonium fission. This can vary from 30 to 70% depending on the design and working conditions of a reactor.

A fast breeder reactor uses plutonium (or highly enriched uranium) as fuel with no moderator. The fast neutrons sustain the chain reaction. Though fast neutrons are less efficient in causing uranium fission, they are ideal for plutonium production because of their greater efficiency in converting U-238 to Pu-239. Also in a reactor using plutonium as its basic fuel, the number of neutrons produced per (Pu-239) fission is 25% more than with U-235. There are thus enough neutrons not only to maintain the chain reaction but also to continually convert U-238 into fissile Pu-239. A metal coolant such as sodium or sodium-potassium alloy acts as a very efficient heat transfer medium without causing any neutron moderation. Thus the plutonium fuel in a fast reactor can produce fission energy as well as breed Pu-239 fuel from the fertile U-238. The 'breeding ratio' (defined as the ratio of fissile nuclei produced to fissioned nuclei) in a fast reactor can exceed 1.0 (i.e. more fuel is created than burnt) under suitable conditions. In contrast, the breeding ratio in a thermal reactor is only 0.6. By optimising the design it is thus possible to generate in a fast reactor as much as 30% more fuel (breeding ratio of 1.3) than it consumes. Such a design is called a Fast Breeder Reactor (FBR). Thus, the FBR nuclear reactor is the only furnace which, while producing energy, generates more fuel than what is consumed. The time required for a breeder reactor to produce enough plutonium to fuel a second reactor is called its doubling time. A doubling time of ten years means that a fast breeder reactor could use the heat of the reaction to produce energy for ten years, and by the end of ten years yield enough fuel to start another reactor.

When the commercial nuclear power plant construction picked up momentum in the 1960s there was a strong expectation in the industry that the following two decades would see accelerated reactor growth. At the same time, there was also an apprehension in some quarters that uranium resources might not keep up with demand leading to shortage and price escalation of uranium. Further, a large number of thermal reactors (graphite and water moderated) were being established at the time in most of the developed countries, along with plans for reprocessing of used nuclear fuel, which projected a situation of plentiful supply of plutonium and could be used in an FBR for power generation. Scientists and technologists set about to exploit this potential of an FBR in meeting the long-term world energy needs. Several countries planned to set up FBRs using plutonium as fuel and depleted uranium (fertile) as blanket material for its conversion to Pu-239 by the surplus neutrons. It was even estimated that using the FBRs it would be possible to utilise natural uranium about sixty times more effectively than in a thermal reactor, thus effectively extending the availability of the world's uranium resources.

Most of the early experiments on fast reactors were conducted in the US. The first was with a tiny unit called Clementine built in Los Alamos in 1946. The fuel was plutonium, cylindrical in shape 6" high and 6" across with mercury as the coolant. In 1951 another Experimental Breeder Reactor, EBR-I, was constructed in Idaho. This unit used molten sodium-potassium alloy as a coolant and generated enough electricity to light its own building. EBR-II, which was started in 1959 ran for 30 years with a power output of 300kw. The fuel consisted of enriched (67% U-235) uranium metal rods contained in thin-walled stainless steel tubes. Altogether, the core contained about 300kg of highly enriched uranium fuel called the 'driver.' Outside the driver was a collection of hexagonal S.S. bars surrounded by a blanket of depleted uranium, which undergoes conversion to Pu-239. The arrangement has to be as compact as possible to prevent high neutron leakage out of the core. The heat extracted by the liquid sodium coolant was transferred to a secondary cooling system in which water is used to generate steam for running a turbine. The Pu-239 produced in the blanket was recovered by reprocessing. This design of core and blanket combination (Fig.6) became generally adopted for future breeder reactors. In many cases, the enriched uranium core was substituted by plutonium which offered a better breeding ratio.

The first US commercial FBR, Fermi-I, was commissioned in 1960 in Michigan. With an output of 66 MWe, this reactor operated just for three years before it suffered fuel meltdown. The FBR programme was discontinued in the wake of a policy decision taken by President Carter not to reprocess used nuclear fuel for plutonium extraction from spent fuel elements. The Clinch River Breeder Project was abandoned in 1981 after incurring an expenditure of nearly $1.1 billion.

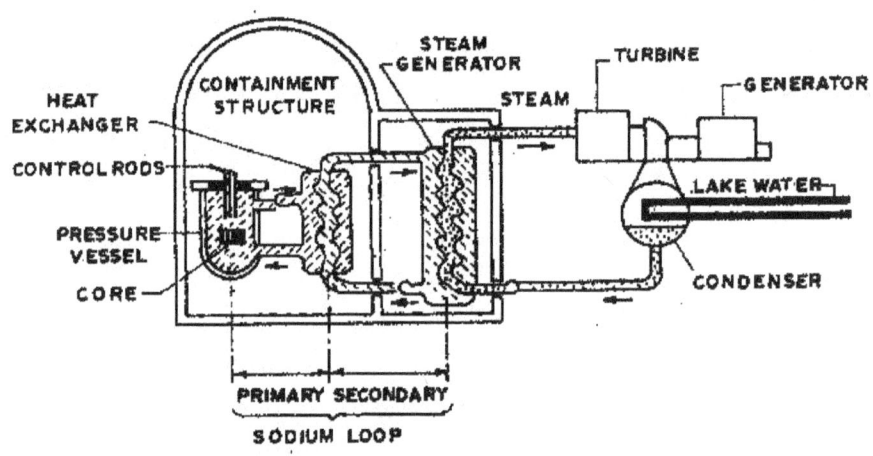

Fig. 6: Liquid Metal-cooled Fast Breeder Reactor (LMFBR)

In 1959, the British Atomic Energy Authority's prototype fast breeder reactor was commissioned at Dounreay. This reactor with a projected output of 15 MWe, faced a number of problems. It was abandoned in 1977. Another Prototype Fast Reactor with an output of 250 MWe achieved criticality in 1974 but was shut down in 1994. The French 250MW prototype fast breeder, Phoenix, went critical in 1974 in France. This reactor, in spite of some problems, did not do badly. Based on its 10-year performance, the French authorities proceeded to build a commercial plant, Super-Phoenix. This 1,200 MWe reactor, which started delivering power in 1986, was the biggest FBR ever operated anywhere in the world. Due to several problems, the reactor operated only for the equivalent of 278 days of full power and was closed in 1997. A breeder built in West Germany, SN-300, was decommissioned without ever having been operated. In the former USSR the BN-350 (130 MWe) reactor operated between 1972 and 1999. Its successor BN-600 (600MWe) began operation in 1980 near Beloyarsk, Russia. In Japan, the 280 MWe reactor at Monju began supplying power in 1994 but was shut down eight months later in December 1995 due to technical problems. In China, an Advanced experimental fast breeder reactor of 60 MWt output achieved full power operation on March 1, 2012.

India's fast breeder programme will be described in the next chapter.

The FBRs with a 'fertile blanket' of depleted uranium around the core set up in different countries have provided an experience covering 400 reactor years. These reactors have been shown to have desirable characteristics such as simplicity, stability in control and the possibility of attaining high energy output and high fuel burn-up (i.e. the fuel can remain active in a reactor for a long time before a replacement is required). Mixed plutonium and uranium oxide fuel (MOX) and sodium coolant are also considered suitable for the designs. In spite of so many positive factors, there are apparently no plans for building reactors with this design in the face of some negative factors. These are:

- The construction and operation of the reactors are prohibitively expensive,
- There are proliferation risks from separation of plutonium required as start-up fuel for breeder reactor,
- The reactors already in operation have been dogged with frequent break-downs,
- The reactors are not economically viable as long as natural uranium is available at reasonable cost,
- The cost of nuclear fuel reprocessing for plutonium recovery is high.

7.6 Generation III Reactors

For more than five decades the nuclear power industry has been improving reactor technology in terms of safety, economy and other considerations. The experience gained with the reactors built and operated so far is now made use of for the construction of a new generation (Generations III and III+) of nuclear power plants with features such as,

- A standardised design for each type to reduce capital costs and construction time and expedite licensing process,
- Simpler and more rugged design for easy operation,
- Longer operating life, typically 60 years as against 40 years for current models,
- Reduced possibility of core melt accidents,
- Lighter burn up to reduce fuel use and down-time.and
- Burnable neutron poisons.

The designs contain several safety features. In the event of severe accidents, safety systems use natural forces such as gravity, circulation and evaporation rather than active systems such as pumps, motors and valve to arrest reactor activity.

The first few of these reactors are under operation in Japan with others under construction. While some of these are evolutionary, mainly from the PWR, BWR and CANDU designs, some represent more radical designs.

7.6.1 Advanced Water Cooled Reactors

The European Pressurized Water Reactor (EPR) is developed by French-German collaboration. With a 1,750 MWe capacity, this reactor is capable of using a full load of MOX (Mixed Oxides of plutonium and uranium) fuel. It also has a 60-year operating life and 92% plant availability. Its optimised core design and higher overall efficiency are expected to result in savings on uranium consumption. From a safety point of view the reactor is claimed to ensure a drastic reduction in the probability of severe accidents and their effects on the environments. In addition, the reactor claimed to be particularly resistant to external accidents such as airplane crashes. The first EPR unit is being built in Finland and the second in France in Flamaville. Two units are under construction in China. France is expecting to build at least one reactor of this type per year starting from 2015 to replace its fleet of 58 existing reactors.

Two units of Advanced Boiling Water (ABWR) unit derived from General Electric with an output of 1212 MWe each have been in commercial operation in Japan since 1996. Negotiations with France are at an advanced stage for the construction of six EPR reactor units (Areva) each of 1650 MWe at Jaitapur, Maharashtra, India.

The Westinghouse Company, in collaboration with the US Department of Energy (DOE) developed an advanced model, called AP 1000 with 1117 MWe output. The first units are under construction in China.

Mitsubishi in collaboration with Westinghouse designed an Advanced Pressurized Water Reactor with an output of 1538 MWe. Two units are coming up in Japan. This design will be the basis of an APWR+ design with 1750 MWe output and full MOX core loading.

The Atomic Energy of Canada has developed designs for CANDU ACR1000 (light-water-cooled, heavy-water-moderated) reactors based on its CANDU reactors. Some recent designs, which are operating in China, have some features like higher thermal efficiency, longer fuel life, reduced high-level waste volumes and also efficient burning of MOX fuel and actinides.

Other PWR designs include APR 1400 (South Korea), Atmea 1 (Areva and Mitsubishi), Kerena (Siemens), AES-92, V-392, AES-2006 and MIR-1200 (Russia) and IRIS (Westinghouse).

India is developing an Advanced Heavy Water Reactor (AHWR) as part of its programme to utilise thorium to fuel its nuclear programme (see next Chapter).

7.6.2 High Temperature Gas Cooled Reactors

High-temperature gas-cooled reactors (HTGRs or simply HTRs) are distinguished from other gas-cooled reactors by the higher temperatures attained within the reactor. Such higher temperatures permit the reactor to be used as an industrial heat source in addition to generating electricity and also for various other purposes. These reactors use helium as coolant, which, at up to 9500 C drives a gas turbine for power generation and a compressor to return the gas to the reactor core. Being chemically inert, helium does not react with the fuel and structural materials, which makes it extremely safe at high temperatures and the best coolant to remove the dense heat flux. The fuel is in the form of TRISO (Tristructural Isotropic) particles less than one mm in diameter. Each particle has a kernel of uranium oxycarbide, with uranium enriched up to 17% U-235. This is surrounded by layers of carbon and silicon carbide which is stable up to 16000°C or more, and acts as a containment for the fission products. These particles are arranged as hexagonal prisms of graphite, or in billiard ball-sized pebbles of graphite encased in silicon carbide.

The best example for the HTR design is the Pebble Bed Modular Reactor (PBMR) being developed by Eskom in South Africa in collaboration with Germany. Each modular unit will have an output of 165 MWe with 42% thermal efficiency. Up to 450,000 fuel pebbles recycle through the reactor continuously until they are expended. The expended fuel pebbles are replaced on line. The pressure vessel is lined with graphite and with a central column of graphite as a reflector. The PBMRs are operated above 250°C, the annealing temperature of graphite, so that Wigner energy is not accumulated, solving the problem that led to the Windscale reactor accident in 1956. Control rods are located in the side reflectors. A demonstration plant is under construction and the fuel loading is scheduled for 2013.

Recent reports, however, say that South Africa is abandoning its plans to develop these reactors. Several other countries including the US, Germany, France, and the UK have also put in serious effort over the last 50 years to develop the technology, but were not able to produce a commercially-viable design. China, which at one stage actively pursued the programme, is also reported to be going slow.

7.7 Generation IV Reactors

International efforts are currently directed to Generation IV programmes. These are of advanced design in which both plutonium and other actinide production and consumption occur in the core itself. These reactors minimise the risk that the spent fuel from energy production would be used for weapons production and also provide the ability to squeeze the maximum energy out of the nuclear fuel.

An international task Force called the Generation IV International Forum (GIF) led by the US and comprising Argentina, Brazil, Canada, China, France, Japan, Russia, South Africa, South Korea, Switzerland and the UK has taken up the task of developing the Generation IV nuclear power reactors with primary focus on electricity generation and also for hydrogen production and nuclear waste treatment.

The GIF task force has set before itself the task of identifying and selecting six nuclear energy systems for further development. These systems employ a variety of reactor, energy conversion, and fuel cycle technologies. Their designs feature thermal and fast neutron reactors, closed and open fuel cycles and a wide range of reactor sizes from very small to very large. Depending on the respective degrees of technical maturity, these Generation IV reactor systems, which are largely based on Fast Neutron Reactor systems, are expected to become available for commercial production after several decades. Six nuclear reactor systems considered suitable for the purpose are,

- Gas-cooled Fast Reactor System
- Lead-cooled Fast Reactor System
- Molten Salt Reactor System
- Supercritical-Water cooled Reactor System
- Sodium-cooled Fast Reactor System
- Very High Temperature Reactor System

The IAEA also has launched a project called Project Innovative Nuclear Reactors and Fuel Cycles (INPRO) to ensure that nuclear energy is available to meet the world energy needs in 21st century in a sustainable manner and to bring together both technology holders and technology users to consider jointly the international and national actions required to achieve desired innovations in nuclear reactors and fuel cycles.

Technology has still a long way to go in its quest for the "ultimate reactor" that can provide solutions to the world energy needs.

8
India's Nuclear Power Reactor Programme

8.1 Introduction

As early as 1944, H.J. Bhabha. the architect of India's nuclear energy programme, declared that "when nuclear energy has been successfully applied for production, say a couple of decades from now, India will not have to look abroad for its experts but will find them ready at hand." His dedicated efforts to promote nuclear energy programmes to this end led the Government of India to create the Atomic Energy Commission under his chairmanship in 1948. This was followed, in 1954, by the creation of the Department of Atomic Energy, through an act of the Parliament, for the execution of policies enunciated by the Atomic Energy Commission. During the same year, a multidisciplinary centre for research and development called the Atomic Energy Establishment Trombay (AEET) was set up in Trombay, Mumbai (then Bombay). This centre was renamed the Bhabha Atomic Research Centre (BARC) after the tragic death of Dr. Bhabha, in an air accident in January 1966. Ever since its inception, the Centre directed its efforts to develop scientific and technological ability indigenously in designing and executing projects largely through its own efforts, overcoming several hurdles.

8.2 Nuclear Reactors for Research and Development

The first reactor project taken up by the Centre was the building of an experimental research reactor APSARA, a 1 MWt (Megawatt thermal) capacity swimming pool type reactor. Using six kilograms of enriched uranium fuel rods supplied by the UK Atomic Energy Authority, the entire system was assembled by indigenous efforts. The reactor, which became critical in 1956, proved very useful for the production of isotopes as well as providing good training to the scientific personnel. Under the recent Indo-US nuclear deal, the reactor was shut down in 2010 and the fuel of the reactor was shifted from BARC and put under safeguard.

The construction of the Canada-India-Reactor-US (CIRUS), a 40 MWt natural CANDU type uranium-heavy water moderated research reactor began with Canadian collaboration in 1956. The reactor went critical in 1960. The heavy water required for the reactor was supplied by the US. Through active participation in the installation and operation of the reactor, the Indian scientists gained expertise in reactor technology, such as uranium extraction and purification, fuel fabrication, reactor control and instrumentation, reactor construction, and radioisotope separation.

The 1956 Indo-Canadian agreement prohibited the use of plutonium produced in the reactor for non-peaceful purposes. India pledged to the US to use the CIRUS reactor only for "peaceful purposes." The agreements did not, however, include any enforcement mechanism. The reactor was not under IAEA safeguards either because these regulations did not exist when the reactor was sold. This facilitated India to set up a reprocessing plant for extraction of plutonium. With the commissioning of this plant in 1964, India became the fifth country in the world with reprocessing facilities. Interpreting the prohibition clause in the agreement to exclude "peaceful nuclear explosions" and using the plutonium extracted, India conducted its underground nuclear test at Pokhran in 1974. This event marked the cessation of all nuclear cooperation from Canada and the US. Canada stopped all fuel shipments. The Indian scientists and engineers developed indigenously the technology for operating the reactor. After serving for 50 years, the CIRUS reactor was shut down permanently in December 2010 under the Indo-US nuclear accord. The Atomic Energy Commission announced that as a replacement for the CIRUS reactor, a multipurpose research reactor will be built in the new campus at Visakhapatnam, Andhra Pradesh.

Another landmark achievement for India was the commissioning the DHRUVA Reactor, with a rated output of 100 MWt. This natural uranium heavy water moderated reactor, similar to the CIRUS, was designed and built completely with indigenous expertise and became fully operational in 1985. The plutonium (about 30 kg/year) extracted from the spent fuels of the two reactors CIRUS and DHRUVA provided the material for the Indian scientists to make notable progress in reactor technology and other strategic programmes.

8.2.1 Experimental Reactor Assemblies

Several critical reactor assemblies were also undertaken by Indian scientific personnel for collecting information on design, construction, design, and operation of future nuclear reactors.

The Zero Energy Reactor for Lattice Investigations and New Assemblies (ZERLINA) assembled at BARC was a tank type assembly with natural uranium in the form of rods as fuel and heavy water as moderator and coolant. Producing a nominal power of 100 watts, the reactor went critical in 1961. It was decommissioned in 1983.

PURNIMA-I was the first experimental fast reactor assembly, which became operational in 1972. This tank type of reactor had 21.6 kg of Plutonium Oxide ($PuO2$) in the form of pellets. It produced a power of 1 watt. The reactor, which provided information on the chain-reacting Plutonium system, was operated till 1974.

PURNIMA-II was a tank type homogenous critical assembly containing 440 grams of Uranium-233 in the form of uranyl nitrate solution as fuel with the aqueous component as the moderator. Beryllium oxide was used as a neutron reflector. Reaching criticality in 1984, it produced a power of 10 watts. The reactor was helpful in understanding the feasibility of uranium-233 fueled reactors.

PURNIMA-III contained uranium-233-aluminium alloy as the fuel. It attained criticality in 1990. Decommissioned in 1991, the core of PURNIMA III later became the core of the KAMINI reactor assembly at the Indira Gandhi Centre for Atomic Research (IGCAR) at Kalpakkam (see later section).

8.3 Commercial Power Reactors

The first commercial nuclear power reactors in India were the BWRs (TAPS-1 and TAPS-2) located in Tarapur, about 100km from Mumbai. These reactors were set up on a turn-key basis by General Electric Co., of USA and Bechtel Corporation, following an agreement between the governments of the US, India, and IAEA. The reactors were commissioned in 1969. The original design was for producing 200 MWe from each of the reactors. Due to problems encountered over the period the rated capacity of each reactor is presently downgraded to 160 MWe.

Though the reactors were designed to run on low-enriched uranium oxide fuel, the Indian scientists and engineers successfully ran the reactors for some time with mixed uranium-plutonium oxide (MOX) fuel. This was necessitated when the US went back on the agreement to supply enriched uranium fuel in the wake of the Pokhran test in 1974. Supplies of such fuel could not be easily obtained from other sources like France, Russia, and China. In recent years, Russian supplies have been restarted and the reactors are running with enriched uranium fuel. Two additional reactors (Tarapur 3 and 4) of PHWR type with a power output of 540 MWe each have been commissioned during 2005–2006.

While the Tarapur project was taking shape, efforts began to set up a pressurised natural uranium-heavy water reactor (PHWR) power plant at Kota in Rajasthan. Consisting of two 220 MWe reactors: RAPS-1 and RAPS-2 (Rana Pratap Sagar Project), this was a collaborative venture with Canada, with India retaining the responsibility for construction and installation based on the design and equipment supplied by Canada. For India these two PHWRs were more important than the BWRs as they were to serve as prototypes for future reactors in the country for the following reasons:

- Though in the early stages the heavy water requirements were met by imports, it has become possible to develop the technology indigenously for heavy water manufacture.
- The use of natural uranium as fuel with heavy water as a moderator obviates the need for developing expensive fuel enrichment facilities or dependence on enriched uranium fuel on foreign suppliers.
- Heavy water moderator gives high neutron economy resulting in a low uranium fuel requirement.
- Production of fissile plutonium needed for India's reactor development programme can be achieved.
- Fabrication technologies are within the capabilities of indigenous industry.

The technology for heavy water production was developed indigenously to a level where India has now become the world's second largest producer of heavy water. India even started exporting heavy water.

According to a 1995 estimate by Wisconsin, the heavy water units located at Nangal (now closed), Tuticorin, Vadodara, Kota, Talcher, Hazira and Manuguru have a total production capacity of 360 tonnes of heavy water per year. The Heavy Water Board, constituted by the Department of Atomic Energy is primarily responsible for the production of heavy water.

The Nuclear Power Corporation of India, under the Department of Atomic Energy, was established by the Government of India in 1987 with powers to operate the commercial nuclear power stations and undertake new nuclear power projects for commercial generation of electricity.

While the construction of the Rajasthan reactors was in progress, a programme was taken up in 1970 to set up twin reactors (Madras Atomic Power Station: MAPS-1&2) each with a capacity of 220MWe at Kalpakkam in Tamilnadu. The design, fabrication of equipment, construction of the plant and its commissioning was entirely indigenous. Backed by the experience of RAPS, the Power Project Engineering Division (PPED) introduced many original engineering changes in the design of MAPS reactors. These include the use of seawater for cooling required an additional cooling loop and construction of a half-a-kilometer-long tunnel under the sea for drawing sea water to the reactor. Faced with sanctions in the context of the Pokhran test, India took a long time of over 15 years to commission these reactors (1984 and 1986). These units provided the technical personnel very valuable experience and feedback in almost all fields of design, manufacturing, construction, commissioning and operation of reactors and in standardising the design of all future 220 MWe PHWR reactors. The Indian industry also developed the capability to meet the requirements of the nuclear power industry. A number of units of this standardised design are currently in operation. Work is also in progress on the installation of four indigenous 750 MWe PHWR units at Kota and Kakrapar. Table 1 gives details of various power reactors currently under operation.

Table 1: Power Reactors in Operation

Reactor State Output(MWe) Year of commissioning
TAPS 1&2
(BWR) Maharashtra 160 each 1969
TAPS 3&4 Maharashtra 540 each 2006 and 2005
RAPS-1&2 Rajasthan 100 &200 1973 &1981
RAPS 3&4 Rajasthan 220 each 2000
RAPS 5&6 Rajasthan 220 each 2010
MAPS 1&2 Tamilnadu 220 each 1984 &1986
Kaiga 1 &2 Karnataka 220 each 1999–2000
Kaiga 3 do 220 2007

Kaiga 4 do 220 2011		
Narora 1&2 UP 220 each 1991–92		
Kakrapar 1&2 Gujarat 220 each 1993 & 1995		
Narora 1&2 UP 220 each 1991–92		

Source: Nuclear Power Corporation of India (2011).

(Note: TAPS 1 & 2 are BWRs. The rest are all PHWRs)

The above 20 reactors currently have a total power capacity of 4780 MWe contributing about 3% of the country's total energy generation. Some 350 reactor years of operation has been achieved by the end of 2011.

8.4 India's Nuclear Power Production Targets

According to a statement of the Government of India in the Rajya Sabha on December 2, 2010, India has 149,654 tonnes of proven uranium reserves as on October 31, 2010. But these reserves (mere 0.2% of world reserves) represent low-grade ores containing uranium just about 0.12% U, compared to ores with up to 12–14% in certain resources abroad. The extraction process from these sources makes Indian nuclear fuel 2–3 times costlier than international supplies.

One of the hurdles faced by India in its nuclear programme has been the shortage of domestic uranium and this has proved an impediment to India's nuclear energy targets. The shortage has been further exacerbated by India's exclusion from international nuclear trade through sanctions imposed in the wake for the two nuclear tests conducted at Pokhran in 1974 and 1998. As a result, the work for several new reactors had to be postponed while the existing nuclear reactors could be operated only at partial capacity for the optimal use of the available fuel. For example, the capacity factors during 2006–2010 ranged between 53% and 64% against availability factors ranging between 82% and 92%. The recent Indo-US civil nuclear agreement and other agreements that followed have now enabled India to import uranium fuel and boost the reactor power output. The uranium supplies were also augmented by the recent rise in domestic production of uranium. The Nuclear Power Corporation has now set before itself a nuclear power production target of 20,000 MWe by 2020 and 63,000 MWe by 2030. With this ambitious programme, India has now joined the US, France, Japan, Russia, China and Japan the world's elite club generating nuclear power.

8.4.1 Reactors under Construction

Currently, four more reactors are under construction. These are two VVER (BWR) reactors at Kudankulam, Tamil Nadu with Russian aid and are expected to be operational by 2013. Russia will supply all the enriched fuel for the reactor while India will reprocess the spent fuel and keep the plutonium.

RAPS 7&8 (PHWR) (700 MWe each) of indigenous design to be operational by 2016.

Kakrapar – 3&4 (PHWR) (700 MWe each) of indigenous design to be operational by 2015.

8.4.2 Reactors Planned and Proposed

The Nuclear Power Corporation has planned or firmly proposed several power reactors. Table 2 contains the list of reactors planned.

Table 2: Nuclear Power Plants Planned

Reactor State Type MWe (gross)
Kudankulam 3 & 4 Tamilnadu PWR-AES 92x2 2,100
Jaitapur 1 & 2 Maharashtra PWR-APR x2 3,400
Kaiga 5 & 6 Karnataka PHWR x2 1,400
Kudankulam 5 & 6 Tamilnadu PWR-AES 92 or AES 2006 2,100–2,400
Gorakhpur/Kumbharia Haryana PHWR x 2 1,400 1 & 2
Chutka 1 & 2 MP PHWR x 2 1,400
Bheempur 1 & 2 MP PHWR x 2 1,400
Banswada Rajasthan PHWR x 2 1,400
Kalpakkam 2 & 3 Tamilnadu FBR x 2 1,000
Total 18 units 15,300–15,900

Source: WNA updated September 2012

Note: All the units except the Kalpakkam FBR units will be built by NPCIL.

The Kalpakkam units will be built by BHAVINI (Bharatiya Nabhikiya Vidyut Nigam)

In addition, about 39 more units with a total output of about 45,000 MWe are firmly proposed (Table 3).

8.5 Nuclear Power Reactors for Export

In September 2010, the Chairman of the Indian Atomic Energy Commission announced at the 54th General Conference of the IAEA that the Nuclear Power Corporation of India is "ready to offer Indian

PHWRs of 220 MWe or 540 MWe for export." He further said that the Indian industry is also "on the way" to becoming a competitive supplier of special steels, large-scale forgings, control instruments, software and other nuclear components and services to the global market.

Table 3: Nuclear Power Plants Proposed

Reactor State Type
Kudankulam 7 & 8 Tamilnadu PWR-AES 92 x 2
Or AES-2006
Kumharia/Gorakhpur
3 & 4 Haryana PHWR x 2
Rajauli Bihar PHWR x 2
? PWR x 2
Jaitapur 3 & 4 Maharashtra PWR-EPR x 2
? FBR x 2
AHWR
Jaitapur 5 & 6 " PWR-EPR
Markandi Orissa PWR
Mithi Virdi (1–4) Gujarat AP-1000 x 4
Kovvada (1–4) AP ESBWR x 2
Nizampatnam 1–6 AP ? x 6
Haripur 1 & 2 WB   PWR VVER-1200 x 4
(to be relocated)
Haripur 3 & 4 WB PWR VVER-1200 x 4
Pulivendula AP PWR or PHWR?
Chutka 3 & 4 MP PHWR x 2
Mithi Virdi 5 & 6 Gujarat AP-1000 x 2
Kovvada 5 & 6 AP ESBWR x 2
ESBWR – Economic Simplified Boiling Water Reactor designed by Hitachi

Source: WNA, updated September 2012

8.6 Thorium for Nuclear Power Generation

Thorium, a fertile element and a potentially attractive energy source, is three times more abundant than uranium. India has the second largest thorium reserves in the world. (Table 4).

Table 4: World Thorium Reserves

Country	Economically Extractable Reserves (tons)
Australia	300,000
India	290,000
Norway	170,000
USA	160,000
Canada	100,000
South Africa	35,000
Brazil	16,000
Malaysia	4,500
World Total	**1,200,000**

Source; US Geological Survey, 2005

Thorium has a single isotope, Th-232, which is not fissionable by itself. However, it is a fertile isotope, which gets converted to fissionable U-233 when exposed to slow neutrons according to the reaction,

$$_{90}Th^{232} + {}_0n^1 \rightarrow {}_{90}Th^{233} \rightarrow {}_{91}Pa^{233} + {}_{-1}e^0 \rightarrow {}_{90}U^{233} + {}_{-1}e^0$$

$$(23.5 \text{ min}) \; (27.4 \text{ d}) \; (1.59 \times 10^5 \text{ y})$$

The U-233 thus, generated has a long half-life of 1.592×10^5 years and is as good a nuclear fuel as U-235 and Pu-239 and in some respects even better. This observation created interest in the study of nuclear properties of both Th-232 and U-233 in the 1950s. The fact that U-233 nucleus releases more than two neutrons per fission also offers scope for U-233 fuelled reactors to breed this isotope from Th-232 used as fertile blanket material. Jumpstarting with a fissile material like U-235 or Pu-239, a breeding cycle similar to, but more efficient than that involving U-238 and Pu-239 can be set up. The irradiated fuel can then be processed for separation of uranium-233 fuel, which can then be fed into another reactor as part of a closed fuel cycle (Fig 1). The intermediate product protactinium-233, however, is a neutron absorber which diminishes U-233 yield.

The importance of thorium, which is abundantly available in India, as a fertile material for generating nuclear electricity in the country through a three-stage programme was highlighted at the Second Geneva Conference (1958) by H.J. Bhabha. The first stage involves the construction of power reactors using natural uranium as fuel. In the second stage, the plutonium generated in the first stage would be the main fuel in fast breeder reactors, using uranium or thorium as the blanket material. The U-233 generated from Th-232 in the second stage reactors will be the main fuel for the third stage.

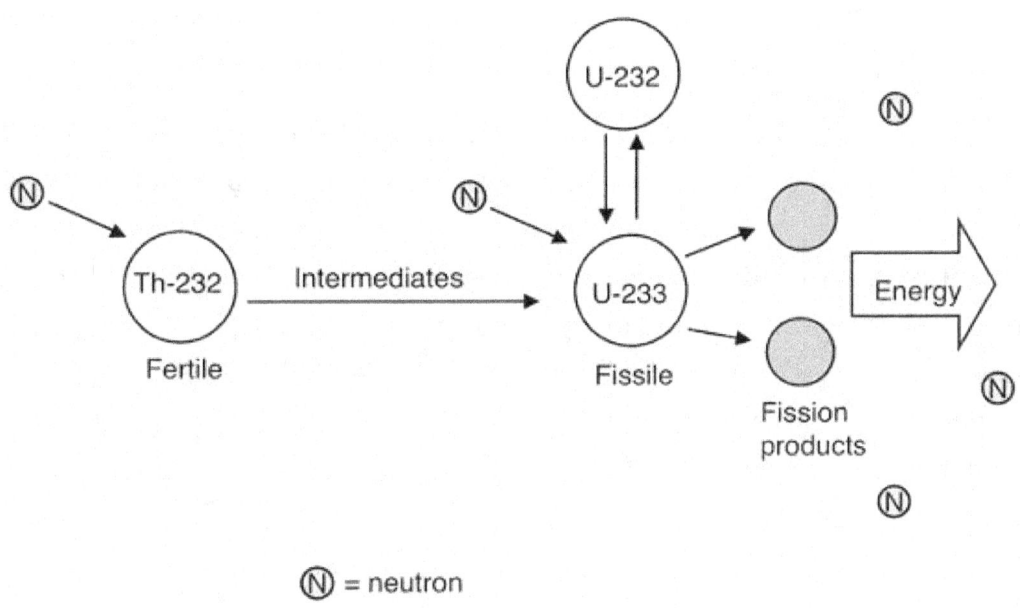

Fig. 1: Thorium Fuel Cycle

(Note: the intermediate stage of Pa-233 is not shown)

8.6.1 International Experience with Thorium Based Fuels

Actual efforts have been ongoing over the past 40 years to utilise thorium as a breeder material. One of the earliest attempts to test the Th-U (233) system was made in the PWR at Shippingport, US. Operated from 1977 to 1982, this Light Water Breeder Reactor was fuelled with Th-U(233). When the reactor was closed, it was found that it contained 1.39% more fissile material than the amount loaded into it, providing experimental evidence for breeding of U-233 in the system.

Basic research on thorium-based fuels has also been conducted in Germany, Japan, Russia, the UK and the USA where several test reactors have either been partially or completely loaded with thorium.

Between 1967 and 1998 an experimental Pebble Bed Reactor at Jülich, Germany, using Th-HEU oxide fuel operated for over 750 weeks with a 15MWe output. Irradiation tests of Th-HEU fuel elements were also made in Dragon Reactor (20MWth) at Wilfrith, UK. A 'breed and feed' regime was followed in which the U-233 formed was separated and used in place of the U-235. In the Netherlands, an aqueous homogenous reactor with Th-HEU fuel was operated for three years. Through continuous reprocessing to remove fission products, a high conversion rate to U-233 was achieved. The high-temperature gas cooled reactor at Peach Bottom in the US was operated for seven years with Th-U 233 fuel.

Large reactors with thorium fuels were also operated for limited periods. A 300MWe high-temperature Pebble Bed type reactor was operated in Germany with Th-HEU fuel during 1983–88. Similar tests were carried out in the US too.

These experiments established that a variety of standard type slow neutron reactors could also be adapted for running with thorium-based fuels.

An advanced concept of a Molten Salt Breeder Reactor (MSBR) was run at Oak Ridge National Laboratory, US, in the 1960s. The fuel was a molten mixture of metal fluorides in which enriched uranium, thorium or U-233 fluoride is dissolved. The fission products dissolve in the salt and are removed continuously in an on-line reprocessing loop. The protactinium formed as an intermediate product is continuously removed from the fuel preventing its conversion to uranium-232 in the reactor, thus considerably reducing one of the major problems with Th-U233 breeders. The project was however stalled in 1969. There is now a renewed interest in the MSBR concept in Japan, Russia, France, and the US. This reactor concept is one of the designs chosen for development in a futuristic plan for Generation IV reactor systems.

8.7 Indian Programme

With the objective of achieving energy security by utilising its vast thorium resources, India initiated a programme to develop fast breeder reactor technology. As part of R&D activities in this direction a test reactor, Kalpakkam Mini Reactor (KAMINI), using uranium-233 fuel produced from thorium irradiated in other reactors, was built and commissioned in 1996 at Kalpakkam, Tamil Nadu. The reactor fuel is an alloy of U-233 and aluminium in the form of flat plates, assembled in an aluminium casing. Demineralised light water was used as moderator, coolant as well as a shield. This reactor, which operates with a nominal power of 30 kWt, is used for neutron radiography, and irradiation purposes. KAMINI reactor is unique as it is the first reactor in the world designed specifically to use uranium-233 fuel.

8.7.1 Fast Breeder Reactor

Vikram Sarabhai who succeeded Bhabha as Chairman, Atomic Energy Commission in 1966 entered into an agreement with France to set up an experimental fast breeder reactor, based on the French "Rhapsodie" design, at the reactor R&D Centre at Kalpakkam, Tamil Nadu. In 1985, this centre was named the Indira Gandhi Centre for Atomic Research (IGCAR). This reactor was to be the forerunner to the second stage of the Indian nuclear power programme.

According to the Indo-French agreement, France would help India set up a sodium-cooled Fast Breeder Test Reactor (FBTR) test facility, which could generate 40MWt. In order to gain experience in power production, a steam generator was added to the FBTR. The French were to supply enriched uranium fuel and also special equipment. However, in the wake of the Pokhran I (1974) test, France pulled out of the collaboration agreement. As a result, Indian scientists and engineers had to rely on indigenous efforts and supplies for completing the project. Instead of enriched uranium fuel, which was to be supplied by France, they decided to use mixed carbides of plutonium and uranium as fuel. Sodium was obtained from a local source and purified to reactor grade. With a few design modifications, the FBTR finally achieved criticality in October 1985. Like the others in the world, this FBTR too threw up a number of challenges. On the whole, the performance of the reactor over its 25-year period has been satisfactory.

During this period the reactor has operated cumulatively for more than 29,000 hours (with about 15,000 hours at a high output of 17.4 MWt (2.5MWe), generating more than 2 million units of electricity. More importantly, the reactor has served as a test bed for fast reactor fuels and structural materials. Efforts are on to achieve a nominal power level of 40 MWt through expanding the fuel assembly in a phased manner.

Encouraged by the satisfactory performance of FBTR, India launched on the construction of a 500 MWe Prototype Fast Breeder Reactor (PFBR) at Kalpakkam. The reactor is a sodium-cooled, mixed oxide (MOX) fuelled pool-type fast reactor. It is designed to have the flexibility to function with metallic fuel, which would give a higher breeding ratio compared to oxide or carbide fuels. Enriched boron carbide will be used as neutron absorber material for reactor control and shutdown. As a part of India's plan to step up nuclear power to 20,000MWe by the year 2020, four additional FBRs, each of 500MWe capacity are also planned. The design and construction of FBRs are entrusted to Bharatiya Nabhikiya Vidyut Nigam Ltd. (BHAVINI), an organisation under the control of DAE, incorporated in 2003. These fast breeder reactor projects would take the country firmly into the second stage of nuclear power strategy.

8.7.2 Advanced Heavy Water Reactor (AHWR)

The Advanced Heavy Water Reactor (AHWR), a generation III design, is also being developed by India. The reactor which bears many features of the PHWR aims to meet the objective of using thorium fuel cycles for commercial power generation. It is not a breeder reactor, but conserves significant quantities of uranium (as fuel). Thorium, used as feed, is converted to fissile U-233, which is burnt in the reactor itself to generate energy. Plutonium is added as an external feed to make the reactor critical. The AHWR is a vertical pressure tube-type of reactor cooled by boiling water under natural circulation with several in-built safety features. Mixed plutonium-thorium oxide and thorium-uranium oxide are used in different parts of the same fuel cluster. Two moderators, amorphous carbon in the fuel bundles and heavy water outside, are used. The fuel cluster consists of 54 pins that are 3.5m in length in three concentric circles surrounding a central displacer rod. The outer circle contains 24 thorium oxide-plutonium oxide fuel pins while the inner two circles contain 30 (total) thorium oxide-uranium-233 oxide fuel pins. The control rod carries dysprosium oxide in a zirconium oxide matrix. The fuel is designed for an average burn-up of 24GWd/tonne.

One notable disadvantage in using thorium is the formation of U-232 along with U-233. This isotope, which has a half-life of 68.9 years, is formed in small quantities from U-233 through (n, 2n) interaction. Its daughter products are hard gamma emitters with short half-lives. Though formed in small quantities, the U-232 entity along with its daughter products contributes a larger dose than U-233. The U-232 in the irradiated material also poses problems during fuel fabrication (high costs) and reuse. This high gamma activity of U-232 is, however, considered an advantage on considerations of non-proliferation.

While initially using plutonium as the external fissile feed, the design aims at making the system self-sufficient in U-233 after initial loading. The spent fuel will be reprocessed and thorium and U-233 will then be recycled and reused. The reactor also incorporates a number of passive safety features as well as a fuel cycle with reduced environmental impact. Several other features also help reduce capital and operating costs. Plans are afoot to set up a prototype reactor with 300MWe capacity.

8.8 Novel Reactor Concepts

Twin Breeder Reactor: Scientists at the BARC have proposed a new concept for a thorium breeder reactor with no feed enrichment. They have shown theoretically that a design, using a judicious mix of the 'seed' plutonium and fertile thorium and uranium zones inside the core, can breed not one but two nuclear fuels (U-233 and Pu-239) within the same reactor and provide sustained energy. For this reason, they call the reactor a Fast Twin Breeder Reactor (FTBR). The design exploits the fact that U-233 is a better fissile material than plutonium. Secondly, the breeding is maximised by putting the fertile materials in an appropriate geometry inside the core rather than as a 'blanket' surrounding the core as is done normally. According to their calculations, such a sodium-cooled FTBR, while consuming 10.96 tonnes of plutonium to generate 1000 MWe, breeds 11.44 tonnes of plutonium and 0.88 tonnes of U-233 in a cycle period of two years. The concept has attracted considerable attention from the scientific world.

Compact High Temperature Reactor: Yet another concept is the Compact High Temperature Reactor. This reactor is mainly a U^{233} – Thorium fuelled, lead-bismuth cooled and beryllium oxide moderated reactor. Initially developed to generate about 100kWt power, this reactor will have a core life of 15 years with several advanced safety features to enable its operation as a compact power unit in remote areas not connected with a power grid. It is designed to operate at 1000^0C to facilitate the demonstration of applications such as hydrogen production through splitting water thermochemically. The efficiency of the thermochemical splitting varies between 40–57%. By contrast, the efficiency of the electrolysis is about 27%.

With the expertise gained in fast reactors and thorium fuel cycle, India has the potential to become a world leader in this area in the coming decades.

9

Spent Nuclear Fuel Reprocessing

9.1 Introduction

Nuclear fuel reprocessing is chemical (or metallurgical) processing of nuclear fuel after its use in a nuclear reactor. Under the US Manhattan Project, Hanford and Oak Ridge were chosen to locate the world's first large nuclear reactors for generating plutonium and reprocessing the irradiated fuel to separate the plutonium. The plutonium separated went into the atom bomb 'Trinity' tested at New Mexico on July 15, 1945, and the bomb 'Fat Man' that was dropped over Nagasaki on August 9, 1945.

The post-World War period saw the Soviet Union, UK and France produce plutonium for nuclear weapons through the chemical reprocessing of nuclear fuel from reactors built for the purpose. Thus, chemical reprocessing became an integral part of the nuclear weapons programme.

The post-war period also saw the use of nuclear energy for commercial power generation. Chemical fuel reprocessing technology enabled the recovery of valuable plutonium and residual uranium from the fuel discharged from the reactors and reuse as well as the separation of the highly radioactive fission products for safe disposal.

The projected rapid growth of nuclear power in the 1970s and rising demand for uranium created apprehensions about a possible shortage of uranium and its price escalation. At that point, the development of 'Breeder Technology' opened up the possibility of converting non-fissionable U-238, which constitutes nearly 99.3% of natural uranium into fissionable Pu-239. It was estimated that through such conversion the energy available from a given quantity of mined uranium could be increased multifold. Attracted by this idea, the industrialised nations started the construction of breeder reactors. The plutonium required as the start-up material for these reactors was extracted from the spent fuel of the PWR and PHWR reactors. However, with the projected growth of nuclear power and the consequent rise in demand for uranium not materialising, nuclear fuel reprocessing became only an optional proposition in nuclear power production.

9.2 Discovery of Neptunium and Plutonium

The discovery of nuclear fission in 1939 by Hahn and Strassmann through the identification of the fission products can be considered the beginning of the chemical reprocessing of nuclear fuel. This was followed, in 1940, by the isolation of the first transuranium element neptunium by McMillan and Abelson by exposing uranium oxide to slow neutrons. This beta-active species ($_{93}Np^{239}$) had a half-life of 2.3 days. Soon after, Seaborg, McMillan, Kennedy and Wahl synthesised another isotope of neptunium ($_{93}Np^{238}$) by bombarding uranium oxide with fast deuterons from a cyclotron,

$$_{92}U^{238} + _{1}D^2 \rightarrow \, _{93}Np^{238} + 2\,_{0}n^1$$

This beta active product also has a very short half-life of 2.1 days. In view of the short half-lives of the two neptunium isotopes, it was not possible to study the chemistry of this new element at that stage. But with the discovery of the longer-lived alpha-active isotope Np^{237} with a half-life of 2.2×10^6 years by Wahl, and Seaborg in 1942, the chemistry of this element could be studied. This radioactive species too was available in quantities too small (called trace quantities) to precipitate through conventional methods. A technique known as the carrier technique was resorted to for its separation. This technique involves adding an appreciable quantity of a chemical (called carrier) with a similar crystalline structure to the solution to incorporate the desired radioactive species present in quantities as small as 10^{-10} gram or even less. When the carrier is subjected to precipitation the tracer element is also carried with it. After separating radioactive trace elements from the carrier, it is possible to study its properties, provided it has a reasonably strong radioactivity. For manipulating these materials in such low concentrations, it was necessary to develop techniques to work on a scale very much below that of microchemistry. The microchemical technique pioneered by A.A. Benedetti-Pichler (*Introduction to Microtechniques in Inorganic Chemistry*, Wiley, 1942) was scaled down and adopted for the purpose by Paul L. Kirk. This technique has come to be called ultramicrochemistry. The extremely small volumes (from 10^{-5} to 10^{-10} ml) were handled using specially-constructed small capillary containers, pipettes, burettes, micromanipulators, etc. The minute quantities of solids were handled on the mechanical stage of a microscope.

Using this technique, Magnusson and La Chapelle prepared the first sample of neptunium-237 and studied its properties. It was observed that, in solution, neptunium compounds exist in +3, +4, +5 and +6 oxidation states. These variable oxidation states, characteristic of neptunium and other transuranium elements, formed the basis for their separation.

During the 1940–41 winter, Seaborg, McMillan, Kennedy and Wahl observed that an alpha-active product with a half-life of around 90 years was generated from the 2.1 day Np^{238}. This was chemically different from both uranium and neptunium. They showed this was a new new element with atomic number 94 formed by the beta decay of neptunium-238 and named it plutonium. Thus Pu-238 was the first isotopic form of the element plutonium to be identified.

The important isotopic form of plutonium, Pu-239, was characterised by Seaborg, Kennedy, Segre and Wahl in March 1941. Though McMillan and Abelson earlier sought this as a decay product of Np^{239}, they were not successful in establishing its identity as the material was too small. The reaction can be represented as,

$$_{93}Np^{239}\ (t_{1/2} - 2.35\ \text{days}) \rightarrow \, _{94}Pu^{239} + _{-1}e^0$$

The half-life of plutonium-239 was determined as 24,200 years. The most important observation however was that, like U-235, this isotopic species also undergoes fission when exposed to neutrons. Separated chemically from the neutron – irradiated uranium, the element plutonium could serve as a more attractive alternative (compared to U-235 which has to be separated by a more complex process) for use in a nuclear explosive. Using plutonium-238 as a tracer Seaborg and his team of workers developed procedures for the separation of the element using lanthanum fluoride as a carrier. They then proceeded to prepare and isolate the element in larger quantities. For this, about 90 kilogram quantity of uranyl nitrate hexahydrate was bombarded for one to two months with the neutrons produced by bombarding a beryllium target with deuterons at the cyclotron facility of Washington University in St Louis. Through a chain of chemical operations involving solvent extraction, and precipitation with lanthanum as a carrier, Cunningham and Werner prepared the first pure compound of plutonium, plutonium fluoride (PuF_4) that weighed a few micrograms, on August 18, 1942. This was converted into oxide (PuO_2) and was weighed using an ingenious weighing device consisting of a quartz fibre, called the Salvioni balance, on September 10, 1942. One of the marvelous achievements of the Berkeley team of scientists was the unraveling of the chemistry of the two man-made elements, neptunium and plutonium starting with such minute quantities that could not be seen by the naked eye and succeeding in the isolation of visible quantities of the synthetic element, plutonium. In the words of Seaborg, *"These memorable days will go down in scientific history to mark the first sight of a synthetic element, and the first isolation of a weighable amount of any artificially-produced isotope of any element."* This first-ever carrier-free sample of plutonium oxide, weighing 2.77 micrograms, is now displayed in the Seaborg Museum at the University of California, Berkeley.

9.3 Plutonium Production at Manhattan Project

Once the potential of Pu-239 for a nuclear fission was recognised, the task before the scientists was to achieve its large-scale separation from the uranium fuel from the reactors. This task was entrusted to the Metallurgical Laboratory of the University of Chicago (now known as Argonne National Laboratory). Fermi and Szilard, who were already carrying out investigations on the nuclear chain reaction at the Columbia University, also moved to the University of Chicago after their first successful experiment establishing the feasibility of nuclear fission chain reaction through setting up an atomic pile with natural uranium as fuel and graphite as a neutron moderator.

The design of the Chicago reactor was not suited for long-term production of plutonium. A new reactor was therefore designed and constructed at Oak Ridge for producing plutonium in sufficient quantities. Facilities to work out methods for its chemical separation were also set up. The Oak Ridge reactor, which started operating in November 1943, began producing plutonium at the rate of several grams per month by the end of February 1944.

Several methods – precipitation, solvent extraction, volatility, adsorption-elution and pyrometallugy, and pyrochemistry – were considered for the quantitative recovery of the small amounts of plutonium from the large amounts of uranium and fission products. The final choice fell on 'precipitation' because it seemed to offer, at that stage, the greatest certainty of 'at least partial success.' The work of Seaborg and his group at Berkeley at ultramicro scale involving lanthanum fluoride precipitation through an oxidation-reduction cycle was all that was available to build upon for this purpose. It remained to be seen whether this technique could be translated into a production process to extract the plutonium required for the bomb effort. To make the laboratory data as reliable as possible Seaborg's team had to first work with realistic (fairly high) concentrations of uranium and plutonium in solution. For this purpose, the team had only a few hundred micrograms of plutonium at its disposal, which had been isolated by Cunningham and Werner in August 1942 and later in milligram and gram quantities from Oak Ridge. The team faced two major problems in this exercise. One was the enormous radioactivity of the fission products associated with the irradiated uranium. The second was the huge scale-up factor (about 10^9) involved in translating the ultramicrochemical data to the large-scale production level involving tonnes of fuel – an exercise never before attempted in the chemical industry.

One change was incorporated in the precipitation method for plutonium extraction in the fuel reprocessing. Instead of co-precipitation with lanthanum fluoride at the initial stage, co-precipitation with bismuth phosphate was adopted as it offered some advantages. The separation process broadly consisted of the following steps:

- Plutonium was reduced to +4 oxidation state in sulphuric acid. Bismuth nitrate and phosphoric acid were added to the solution to cause the precipitation of bismuth phosphate, which acted as the carrier for plutonium. The precipitate was centrifuged and the clear liquid containing the bulk of uranium and most of the fission products was discarded leaving bismuth phosphate and plutonium as a solid. Plutonium was dissolved in nitric acid by oxidising it selectively to +6 oxidation state. The insoluble bismuth phosphate was then discarded.

- The plutonium in solution was reduced again by adding ferrous ammonium sulphate and the bismuth phosphate precipitation was repeated. The plutonium carried by bismuth phosphate was once again dissolved by oxidising it using permanganate or persulphate.

- The rare earth fission products were separated by co-precipitation with lanthanum fluoride formed by adding hydrofluoric acid and lanthanum nitrate to the solution.

- After reducing plutonium to +4 state, precipitation with lanthanum fluoride was repeated. The plutonium carried was oxidised and brought into solution in a pure form. The purified plutonium was precipitated as hydroxide, separated and dissolved in nitric acid.

- Hydrogen peroxide and ammonium nitrate were added to plutonium solution to precipitate plutonium as peroxide. The precipitate was dissolved in nitric acid. Finally, the plutonium nitrate solution was evaporated to produce a wet paste.

Early in 1943, a site at Hanford in Washington State was chosen for the location of the plant for the large-scale production of plutonium. Du Pont Du Nemours, the biggest chemical company in the US, was commissioned to design, engineer, construct and operate this complex facility. The company was uneasy to take responsibility for this project as they did not have the expertise. They agreed to sign a contract but only with certain conditions. One was that the company would be paid a nominal profit of $1 on a cost-plus-fixed-fee basis. The second condition was that the Government should provide protection against all costs, expenses, claims, and losses because of the hidden hazards in the unconventional job they were executing.

Hanford was the biggest project Du Pont had ever built and operated. Also, this was the first time in human history that mankind was handling such enormous amounts of radioactivity to extract an element that had never existed on the surface of the planet before. Construction work on reactors at Hanford began in June 1943. More than 50,000 workers toiled at this site. Within 30 months, three reactors, three chemical processing plants, and 64 underground storage tanks to store different waste solutions were completed. The number of reactors rose to nine in due course. A fuel fabrication facility and other support facilities were also built. The first reactor, called B reactor, has been recognised as 'a national engineering historic landmark and a nuclear historic landmark.' This reactor, which was completed in just over a year, produced the first plutonium in November 1944. The remotely-operated chemical separation plants called 'canyons' (because of their shape) were located 10 miles away from the reactors. The irradiated fuel slugs were transported from the reactors to the separation plants by rail.

In the words of Seaborg such a scale-up from the ultra-microgram levels to kilogram level operations, *"marks one of the amazing achievements of chemistry."*

The 99.7% pure plutonium, first ever extracted from the nuclear fuel from the Hanford reactors, went into the two plutonium bombs – Trinity and Fat Man.

9.4 Advances in Fuel Reprocessing Technology

With countries in Europe setting up nuclear reactors for weapons development and electricity generation at end of World War II, reprocessing of used fuel became a necessary component of these activities. As the variety and capacity of reactors increased, the need also arose for a more efficient and viable reprocessing technology. The bismuth phosphate process for the separation of plutonium followed at Hanford works had several disadvantages. These include loss of valuable uranium, which goes along with the fission products, generation of large volumes of liquid waste with high salt content and time-consuming batch operations. These disadvantages were overcome by substituting the bismuth phosphate precipitation process with liquid-liquid (solvent extraction) extraction process.

9.4.1 Liquid-Liquid Extraction

The primary advantage of the liquid-liquid extraction process lies in its usefulness over a vide concentration range of solutes (micro to macro). Also when only a limited separation can be achieved in a single, batch equilibrium contact of two phases, the separation can be amplified through an automatic multistage separation process called *countercurrent extraction* which involves repeating single stage contacts between the two liquid phases (organic and aqueous) with the aqueous and organic streams automatically moving in opposite directions.

9.4.1.1 REDOX Process

Ethyl ether and methyl isobutyl ketone solvents are good extractants of uranium. Since ethyl ether is a highly volatile and reactive solvent, methyl isobutyl ketone (hexone) was used in the early extraction processes. This process, known as Reduction-Oxidation (REDOX) process, was first tested in a pilot plant at Oak Ridge in 1951, and later adopted for large-scale operations at Idaho and Hanford. Pu (+6) and U (+6) are extracted by hexone from nitric acid solution with high yield and purity from the fission products. Plutonium in the solvent phase is reduced to Pu (+3) through the addition of ferrous hydroxamate and selectively extracted into the aqueous phase. The extraction cycle is repeated to decontaminate the products. A disadvantage of the process is that a salting-out agent (aluminium nitrate) has to be added to the aqueous phase to achieve reasonable extraction efficiency. Other disadvantages are the low flash point of hexone (60^0C) and its degradation by concentrated nitric acid.

Another extractant β-β'-dibutyl diethyl ether was used at the Windscale (UK) plant till 1976. This process was known as the Butex process.

9.4.1.2 PUREX Process

The use of tributyl phosphate (TBP) as an extractant was developed at the Knolls Atomic Power Laboratory, tested at Oak Ridge and adopted for large-scale operations at the Savannah River Plant in 1954 and at Hanford in 1956. This process, called the PUREX (Plutonium Uranium Reduction EXtraction) process, has since been adopted internationally as the standard method for commercial reprocessing of spent nuclear fuel because of several advantages over Hexone. TBP is less flammable and chemically more stable in nitric acid. The extraction process is also more economical. A 20–50% solution of TBP in a long-chain hydrocarbon (e.g. kerosene) is used as an extractant. The PUREX technology will now be described briefly.

After removing the cladding (aluminium, Magnox, Zircaloy or stainless steel), the spent fuel is dissolved in nitric acid (>0.5 M) at about 100^0 C in the presence of air. The oxygen in the air minimises the loss of nitric acid as nitric oxide.

$$U + 3/2\ O_2 + 2HNO_3 \rightarrow UO_2(NO_3)_2 + H_2O$$

$$UO_2 + HNO_3 + \tfrac{1}{2} O_2 \rightarrow UO_2(NO_3)_2 + H_2O$$

While uranium metal and oxide dissolve readily, plutonium oxide, when present in larger quantities (>20%), causes dissolution problems. If the fuel has undergone high burn up, the so-called noble metal fission products (molybdenum, palladium, rhodium, ruthenium, and technetium) are also formed. These also do not go into solution. The solution is filtered to remove suspended particles. The separation steps that follow are based on the differences in extraction behaviour of uranium and plutonium species in different oxidation states (Table 1).

Table 1: Extraction behaviour of Actinide ions by TBP

Oxidation State	+3	+4	+5	+6
Uranium	Unstable	Extracted	Not Extracted	Extracted
Neptunium	Unstable	Extracted	Not Extracted	Extracted
Plutonium	Not Extracted	Extracted	Not Extracted	Extracted

Note: Fission products are not extracted

The solution, treated with oxides of nitrogen or sodium nitrite to stabilise the Pu(+4) without affecting the U(+6), is fed into the extraction column and extracted with the TBP when Pu(+4) and U(+6) partition into the organic phase and most of the other metals and fission products remain in the aqueous phase (raffinate). To separate plutonium from uranium, a reductant (e.g. ferrous sulphamate) is added to selectively reduce the Pu(+4) to Pu(+3), which goes into the aqueous phase, while U(+6) remains in the solvent phase. The organic phase is separated and partitioned with an aqueous nitric acid solution of low acidity (0.01M) when the uranium is transferred back into the aqueous phase. The uranium and plutonium streams at this stage contain small amounts of radioactive fission products. Another cycle of solvent extraction is adopted for their removal.

Final purification of plutonium is carried out by solvent extraction or anion-exchange. The anion-exchange method is based on the fact that in strong (7M) nitric acid medium Pu(+6) forms an anionic complex which is selectively and strongly adsorbed on an anion exchange resin. The plutonium retained is eluted from the column using dilute nitric acid (0.5M).

The PUREX extraction method has the advantage of continuous operation with a high throughput. When compared to other extraction methods the volume of the highly radioactive and long-lived waste to be disposed of is considerably low. But slow degradation of the TBP solvent occurs due to hydrolysis and radiolysis. The degradation products, dibutyl phosphate, and monobutyl phosphate strongly retain plutonium in the organic phase and make its stripping difficult. These are removed by washing the solvent with a solution of sodium carbonate followed by a wash with nitric acid before reuse.

The separation of the plutonium and uranium from used nuclear fuel by the PUREX process leaves the fission products and minor actinides in the waste solution. This high-level waste is evaporated and subjected to vitrification to form a stable borosilicate glass suitable for long-term storage.

9.5 Modified PUREX Processes

Modifications have been introduced in the PUREX process to recover other long-lived transuranic fissionable components (the Actinides Np, Am and, Cm) from the spent fuel in a proliferation-resistant form. They can also be used in conjunction with the new generation of fast reactors for energy generation. The options available in the modified PUREX reprocessing are,

- Separate U and Pu (as practiced currently)
- Separate U, Pu +U (U in small amounts)
- Separate U, Pu and other minor actinides
- Separate U, Pu+Np, Am+Cm
- Separate U+Pu all together
- Separate U, Pu+Actinides, certain fission products

The US is developing a variety of UREX processes. In the UREX$^+$ process, only uranium is recovered for recycling and the residue is treated to recover plutonium with other transuranics, leaving the fission product waste. The plutonium+transuranics fraction will then be used by recycling in the fast reactors. Several other variations of UREX$^+$ (UREX^{+1a}, UREX^{+3}) are also developed, the differences being in the way how the plutonium is combined with various minor actinides and how the lanthanide and non-lanthanide fission products are combined or separated. Another process called NUEX, separates uranium and then all transuranics (including plutonium) together, leaving behind the fission products.

The Areva plant in France in its COEX process co-extracts the uranium and plutonium (and usually neptunium) instead of the pure individual products. This is then used for making the MOX fuel. COEX may have from 20–80% uranium in the MOX product with a base line of 50%. This is accomplished by adjusting the chemistry of the PUREX process to allow some uranium to be back-extracted from the solvent with plutonium. Other variations include DIAMEX-SANEX and GANEX processes.

The Rokkasho processing plant in Japan uses a modified PUREX process to obtain a 50:50 MOX product.

Mitsubishi and Japan R&D establishments are investigating direct extraction by supercritical fluids. Under this programme, called Super-DIREX, the fuel rods are dissolved in nitric acid-TBP mixture and extracted with supercritical carbon dioxide or Freon at a low temperature. This results in the separation of uranium, plutonium and minor actinides which complex with TBP.

9.6 Pyrometallurgical (non-aqueous) Processes

Pyrometallurgical processes are considered better alternatives to the aqueous processes, especially for the new generation fast reactor systems. The processes produce lesser quantities of effluents. Using these processes, virtually all the transuranics with some uranium and fission products would be separated and converted to metallic fast-reactor fuel for use in the new generation reactors. Another component containing the bulk of depleted uranium would be used as a fast reactor fuel. The third component containing most of the fission products will be sent to a safe disposal site after conversion to a glass-like form. Much of the current state of the art for these processes was developed at the Argonne National Laboratory between 1984 and 1995 as part of the Integral Fast Reactor Programme. About 25 tonnes of spent fuel from the Experimental Breeder Reactor-II, which was shut down in 1995, was used for developing the pyroprocess technology. The Republic of South Korea is currently pursuing this technology for processing fuel from its commercial reactors to minimise the volume of high-level liquid waste and also possibly to extract fissile actinides for fabrication of fast reactor fuel. Russia has already demonstrated the production of MOX based on pyroprocessing and plans to develop a closed fuel cycle using this technology by 2020. The technology is still in the laboratory and at the pilot-scale experimentation stages.

Broadly the pyroprocesses involve dissolving the spent fuel (in the form of metal, oxides or nitrides) in a bath of molten salts (chlorides, fluorides etc) at temperatures of several hundred degrees and then separating the desired species using various techniques such as liquid metal extraction, electrolysis or selective precipitation. These processes have attracted attention mainly because of their ability to dissolve both ionic and refractory compounds, low radiation sensitivity of inorganic salts, thus making them theoretically suitable for 'on-line reprocessing' of the spent fuel from molten salt reactors. Other attractive features are the compact nature of the basic process with very few conversion steps and their suitability for management of the actinides group products. Yet another advantage is their low criticality risk relative to aqueous methods because of the absence of water which acts as an efficient neutron moderator.

9.7 Global Nuclear Energy Partnership

Due to rising worldwide interest in nuclear power production as an alternative to fossil fuel – based thermal power production, several national and international programmes have been launched to develop advanced nuclear fuel cycles. The US has initiated in 2007 the Global Nuclear Energy Partnership (GNEP) with US, France, China, Japan, and Russia as the founding members and IAEA and the European Commission as observers to develop new proliferation-resistant recycling technologies that extract more energy, reduce waste and minimise proliferation concerns. The name of GNEP was changed in 2010 to 'International Framework for Nuclear Energy Cooperation (IFNEC).' So far 25 nations have joined the partnership. This group will be responsible for enriching uranium and manufacturing fuel for power reactors. The above countries called the 'Fuel Cycle States' would lease the nuclear fuel to 'Recipient States' for use in their power reactors. When the fuel is used up in the reactor, the Fuel Cycle State would take back the spent fuel for advanced reprocessing and recovering the plutonium, uranium, and actinides for recycling in advanced reactors. These reprocessing programmes aim at preventing proliferation by keeping the plutonium either with some uranium or with other transuranics which can be destroyed only by burning in a fast neutron reactor. Trials of some fuels arising from UREX+ reprocessing in the US are being conducted in the French Phoenix fast reactor. Other IFNEC programmes include:

- Development of advanced recycling reactor with a capacity of 1000 MWe, basically a fast reactor capable of burning plutonium and other minor transuranics,

- Development of cost competitive fast reactor technology, which would need switch over from aqueous fuel reprocessing to pyrometallurgical processes, and

- Development of methods for the safe disposal of nuclear wastes from recycling operations.

All these activities constitute the IFNEC strategy to develop comprehensive fuel services, including fuel leasing and addressing the challenges of reliable fuel supply to the member nations seeking the benefits of nuclear power without their establishing indigenous fuel cycle facilities. This strategy also aims at reduction of proliferation risks. The US withdrew from the IFNEC following a change in government policy on commercial reprocessing.

9.8 THOREX Process

The successful use of thorium as a breeding source for uranium-233, a nuclear fuel, was first demonstrated in the molten-salt reactor experiment conducted from 1964 to 1969 at the Oak Ridge National laboratory. Thorium fuel cycles are particularly relevant to India, which has large thorium deposits but very limited uranium reserves.

The database and experience of thorium-uranium fuel reprocessing are very limited – as compared to UO_2 and MOX fuels – for commercial exploitation. Most of the experience for the thorium fuel reprocessing has come from the recovery of small amounts of uranium-233 bred in irradiated ThO_2. The first technical-scale separation of irradiated thorium-uranium fuel (THOREX process) dates back to 1952.

The THOREX process is similar to the PUREX process redesigned for reprocessing irradiated thorium fuel. TBP is used in this process also for the separation of U-233 from thorium-232. In the conversion chain of thorium-232 to uranium-233, protactinium-233 is formed as an intermediate. This species has a half-life of about 27 days as compared to neptunium-239 intermediate ($t_{1/2}$ 3.5 days) in the uranium fuel cycle. Consequently, a longer cooling time of at least one year is required for the complete decay of protactinium. The control of protactinium in the separation scheme is a major problem as it could have a long-term radiological impact if it finds its way into the fission products.. It is therefore desirable and even essential to separate protactinium from the spent fuel solution prior to the separation of uranium-233 and thorium. The irradiated Th or Th-based fuels also contain a significant amount of uranium-232, which has a half-life of 73.6 years along with its strong gamma emitting daughter products (Bismuth-212, Thallium-208) with very short half-lives. As a result there will be a significant build-up of radiation dose during the storage of the spent Th-based fuels or separated uranium-233. This necessitates remote and automated reprocessing and fabrication in heavily shielded hot cells. These and other technical factors add significantly to the costs of power generation.

9.9 India's Nuclear Fuel Reprocessing Programme

With low reserves of uranium, reprocessing and recycling of uranium and plutonium was the only choice for India to achieve optimum utilisation of its meagre domestic uranium resources. With no collaboration available, except in the early years, India launched its fuel reprocessing programme way back in 1964 and developed the technology entirely through indigenous efforts. A reprocessing unit with an estimated capacity of 60 tonnes per year was commissioned at Trombay. Since then India has been practicing reprocessing as an integral part of its civilian nuclear energy programme. Two more plants of 100 tonnes capacity each at Tarapur (1975) and Kalpakkam were built to reprocess the spent fuel from the PHWR reactors. A new 100 tonnes per year plant was commissioned at Tarapur in 2011 and another 100 tonne-unit shortly thereafter at Kalpakkam. With 40 years of experience, Indian scientists have gained expertise in reprocessing technology based on the PUREX process.

Russia has agreed to India's proposal to reprocess the spent fuel from the two reactors being set up by the former at Kudankulam. India is also taking steps to secure clearance from France and the US to reprocess the spent nuclear fuel from the power reactors to be set up by these two countries. These units will be under the surveillance of IAEA.

10

Safety in Nuclear Industry

10.1 Introduction

Any undesirable alteration of our surroundings, through direct or indirect effects on energy patterns, radiation levels, chemical and physical constitution and abundance of organisms cause adverse effects on the environment. The adverse effects on the environment from the nuclear energy programmes and the methods adopted for their mitigation will be outlined in this chapter.

X-rays, discovered by Röntgen in 1895, were almost immediately put to use as a diagnostic tool. The acute effects arising from exposure to X-rays were realised in 1896 when Nikola Tesla intentionally subjected his fingers to exposure to these rays.

The discovery of radioactivity, followed by the isolation of radium, set in motion the use of radioactive substances especially radium and radon, for therapeutic purposes. The first radioactive spa began functioning in 1906 at Jáchymov in West Bohemia, Czech Republic. Radium-rich springs became popular in Japan. Drinking therapy (e.g. drinking radon containing waters) was adopted in Germany and inhalation therapy (inhaling the radon from the radon springs and spas) in Austria, Poland, Romania and in the US (Denver). Many physicians began marketing radioactive substances as potent medicines. Madame Curie was the first to warn mankind about the adverse effects of radiation on the human body. Ironically, she died from anemia caused by exposure to radiation from the materials she discovered and handled.

The human population is continually subjected to exposure, both externally and internally, from small amounts of radioactive uranium, thorium and their decay products such as radon present in the rocks and the Earth's crust. The human body contains the radioactive potassium-40 and carbon-14. These sources constitute terrestrial radiation. In addition, people are also exposed to radiation from space. This is called cosmic radiation. The terrestrial and cosmic radiations represent what is known as background radiation. It ranges in magnitude from place to place. For example, people living in granitic zones or mineralised sand zones (e.g. beach sands containing monazite) receive more terrestrial radiation than others, while people living at high altitudes receive more cosmic radiation.

Advancements in nuclear science and technology, and their applications for peaceful purposes such as industry, agriculture, and medicine as well as for military purposes and power generation have further increased the risk of exposure of humans to higher levels of radiation. Starting with the mining of uranium and ending with nuclear power generation all these activities have become sources of radiation in the form of hazardous radioactive effluents and nuclear waste. However, the carefully recorded experiences of pioneer scientists helped in evolving the logistics for the safe handling of hazardous nuclear materials which contributed to the development of nuclear power production without taking a large number of lives.

10.2 Ionising Radiations

Radiation is of two types: *electromagnetic*, and *corpuscular* (or particulate). Electromagnetic radiation comprises gamma rays, X-rays, UV, visible, infra red, microwave and radio waves. These differ only in wavelength, which range from 10^{-11} cm for gamma rays to 10^8 cm for radio waves. The particulate radiation consists essentially of alpha, beta and neutron particles. The highly energetic gamma, X-, and particulate radiations are capable of causing excitation and ionisation in molecules, changing an orbital electron from an atom leaving behind a positively-charged moiety. For this reason these radiations are called ionising radiations. The resulting negatively-charged and positively-charged species which are reactive, give rise to a number of secondary processes in the system. Such processes occur in the constituents of the living cell when the human body is exposed to ionising radiations resulting in their damage. The adverse effects on living systems from exposure to ionising radiations first reported by H.J. Muller in his classic studies published in 1927 led to a concerted study of the problem by international scientific teams. Dr. Muller was awarded the Nobel Prize in 1946 for his contributions.

Exposure of living systems to ionising radiation can occur in two ways: external and internal. When the source of radiation is external to the body, the radiation emanating from the source impinges on the whole body and deposits energy thereon. Damage from external exposure is caused by gamma radiation, energetic beta radiation, and neutrons which penetrate the body. Their effect can be stopped by the removal of these radiation sources or by effectively shielding them.

Internal exposure arises when a living being breathes air contaminated with radioactive substances or consumes water or food containing them. Part of the activity thus, entering the body gets deposited in different organs depending upon the chemical nature of the material. For example, plutonium and strontium get concentrated in the bone, iodine in thyroid and sodium and phosphorus in the whole body, persisting there over a period of time, depending on their biological removal rate. This rate is measured in terms of biological half-life, which is the time required for half of the material initially absorbed by the system to be expelled. The biological half-life values for some radioactive forms of biologically important elements are: sodium-24(whole body) – 19 days, iodine-131(thyroid) – 180 days, phosphorus-32(whole body esp. liver) – 257 days, strontium-90 (bone) 3.9×10^3 days and plutonium-239 (bone) – 7.3×10^4 days. During this period the energy from the radiations is deposited on the tissues at a diminishing rate depending on their half-lives. Alpha particles, which have a higher mass, lower velocity, and greater charge are capable of traveling only a few inches in air and rarely penetrate the outer skin layer of the body. Similarly, soft beta particles also do not normally penetrate beyond the skin. For this reason, sources of alpha radiation and low energy beta radiation (also called soft beta radiation) with maximum beta energy of a few hundred keV, which have short ranges, are hazardous only when they find entry into the human body.

10.3 Biological Effects of Radiation

The human body has not developed an instinctive reaction to exposure to ionising radiations as it has against heat or to some extent against UV light. Consequently, the impact due to radiation at levels above the background radiation will not be felt immediately. The after-effects in the form of symptoms arising from the damage depend upon the type of radiation, the depth of penetration, the extent of the body exposed and the radiation energy absorbed. Further, it also depends on whether the exposure is repeated or prolonged (leading to a cumulative effect) or is received as a single large dose.

Extremely high radiation exposure of 3,000 to 4,000 times the natural background radiation causes a breakdown in body functions due to cell destruction leading to severe disability or even death within a short time. Exposure to high doses of radiation also produces clinically expressed symptoms like nausea, reddening of the skin, and even more acute syndromes. This is also frequently noticed in the case of cancer patients who have gone through radiation therapy. These symptoms can be noticed within short periods of time after the exposure.

In contrast, when the body is exposed to moderate doses the exposed cells are modified rather than killed. Such changes are believed to induce cancers after a certain latency period. When the dose is small, there is a chance that body's repair and defence mechanisms prevent the manifestation of cancer. Exposures to high levels of radiation occur in very rare events such as occupational exposure due to a reactor accident.

The biological effects of radiation can be divided into two categories – somatic and genetic. The somatic effects relate to the radiation damage to the cells, which are concerned with the maintenance of the body functions e.g. blood cells and bone marrow. The somatic effects may further be categorised as 'deterministic and stochastic.' A deterministic effect is one identified as that arising from radiation exposure crossing a particular value (threshold dose). Examples are radiodermatites, malignant growth (cancer), radiation sickness, leukemia, cataract formation or a reduction in life expectancy. Stochastic (nondeterministic) effects are random effects that occur by chance among people who are not exposed to radiation as well as among the people exposed to low levels of radiation. For example, cancer among the normal population is a stochastic effect. According to a WHO report (2013), there were 7.6 million deaths in 2008 due to cancer and 12.7 million new cancer cases in the civilian population. This could be through exposure to radiation or other factors such as smoking and chemicals.

Genetic effects are also stochastic effects. These arise from injury to the genes in the reproductive cells responsible for the propagation of genetic characteristics. The damage to the genes results in sterility or abnormalities in the offspring. There is no threshold level of radiation below which we can say with certainty that stochastic effects like cancer or genetic damage will not occur. Because these effects also occur due to factors other than radiation exposure, their incidence is extrapolated from high doses. Doubling the radiation dose doubles the probability that cancer or genetic effects will occur. Despite intensive research on the health impacts of radiation exposure, the determination of the probability of stochastic effects, especially at low doses and low dose rates, continues to be contentious. The most prudent course, therefore, is to ensure that exposure to radiation follows the principle of 'As Low as Reasonably Achievable (ALARA)' levels.

10.4 Man-made Radiation Sources

Apart from background radiation, we are exposed to several man-made radiation sources, mostly through medical procedures like diagnostics and radiation therapy (e.g. cancer). While the doses from diagnostic X-ray sources are much lower, radiation therapy can reach levels much higher than normal background radiation. But in the therapeutic procedure, the radiation is only targeted at the affected body part under strict medical supervision. Besides medical applications, extremely small amounts of radioactive materials find use in research and teaching institutions. Other areas where exposure to small amounts of radioactive materials can occur are nuclear reactors and their supporting facilities such as uranium mines and mills, nuclear fuel fabrication plants and facilities involved in nuclear weapons as part of the normal operations. These come under the category of 'occupational exposure.'

10.5 Radiation Dose Units

The standard unit of radiation-absorbed dose presently used is Gray (Gy). This corresponds to an average energy deposition of one Joule per kg of the object. Though this represents only a small amount of energy, a Joule of ionising radiation can cause up to trillions of ionisation in the living systems. It is also well-established that the extent of damage to the systems for different types of ionising radiations (α, β etc.) depends on the energy of the radiation and the rate of energy transfer for the same amount of energy deposited. To account for such variations in the radiation dose received by the living systems, the unit Gray is multiplied by a radiation-type dependent quality factor W_R (called Radiation Weighting Factor), and the product is expressed as the *Equivalent Dose* in units of *Sievert* (Sv). Thus, Equivalent Dose (Sv) = Absorbed Dose (Gy) x W_R

The W_R factor is taken as unity for gamma radiation and beta radiation, and rises through neutrons, protons to alpha, fission fragments and heavy nuclei to a value of 20. For gamma radiation with a W_R value of unity which is of widest concern, Gray and Sievert are used synonymously. Since the magnitudes of Gy and Sv are large compared to the exposures normally encountered, their sub-units (mGy, μGray and mSv and μSv) are frequently used.

An older unit of dose measurement that is still in widespread use is REM (Röntgen Equivalent Man). One Sievert equals 100 REMs.

10.6 Radiation Safety Regulations

Since radiation risks may transcend national borders, regulating the safety of people from radiation exposure is the responsibility of every nation. For this reason, international cooperation is necessary to promote and enhance safety globally through an exchange of experience and by improving capabilities to control hazards, prevent accidents, respond to emergencies and mitigate any harmful consequences. To this end, the United Nations Scientific Committee on the Effects of Atomic Radiation (UNSCEAR) was established in 1955 by the United Nations General Assembly with a mandate to assess and report levels and effects of exposure to ionising radiations. The reviews of this body guide the programmes of the international bodies such as the IAEA, ILO, WHO and the International Commission on Radiological Protection (ICRP). The ICRP, the monitoring and regulatory body, prescribes from time to time limits for radiation exposure for the occupational adults and the general public for adoption by the concerned bodies. The Commission's basic philosophy in formulating the standards is:

- No practice involving exposure to radiation should be adopted unless it produces a net benefit to those exposed or to society generally.

- Radiation doses and risks should be kept 'As Low As Reasonably Achievable (ALARA)' levels, economic and social factors being taken into account.

- The exposure of individuals should be subject to dose or risk limits above which the radiation risk would be deemed unacceptable.

The bottom line of these formulations is the firm belief that any level of radiation dose, no matter how low, involves the possibility of risk to human health. At the same time, practical experience shows that small increases over natural levels of exposure are not likely to be harmful, if governed by the above guidelines. The International Atomic Energy Agency (IAEA) is specifically authorised, under the terms of its statute, to implement these safety standards for protection from radiation sources ensuring reduction of risks.

The ICRP in their recommendation ICRP-103 (2007) prescribed that the dose limits for the occupational exposure of any worker shall be so controlled that the following limits are not exceeded,

- An effective dose of 20 mSv in a year averaged over 5 years with not more than 50 mSv in any single year,

- An equivalent dose to the extremities (hands and feet) of 500 mSv a year and to skin 500 mSv a year.

For members of the public that are likely to be exposed to radiation activities, the following dose limits are prescribed:

- An effective dose of 1 mSv in a year and in special circumstances, an effective dose of up to 5 mSv in a single year provided that the average dose over 5 consecutive years does not exceed 1 mSv per year,

- An equivalent dose to the lens of the eye of 15 mSv in a year and an equivalent dose to the skin of 50 mSv per year.

- For pregnant women a dose of 2mSv on the surface of the abdomen and 1 mSv to the foetus during the remainder of pregnancy.

An effective dose takes into account the specific organs and areas of the body that are exposed since all parts of the body are not equally sensitive to the possible effects of radiation such as cancer and mutations. If more than one area is exposed to the total body effective dose is the sum of the effective doses for the exposed area. These limits are adapted to the needs of various countries and enforced by statutory regulatory bodies with a special emphasis on all forms of occupational exposure.

As mentioned earlier, the human system is exposed to radiation continuously from natural radioactive sources. The background radiation at sea level is generally of the order of 1 mSv/y. In certain regions in the world with natural radioactive mineral deposits (e.g. Kerala, Brazil, and China) the value is higher. The contribution from cosmic radiation is significantly low at the sea level. But it rises to 1.7mSv/y at an altitude of 3,000 metres and as high as 44 mSv/y at 18 km above the ground.

10.7 Regulatory Procedures in India

With the commissioning of nuclear reactors and also increase in use of radiation sources outside the Department of Atomic Energy (DAE) for research, medical and industrial purposes, appropriate steps were taken for the protection of workers and the general public from radiation. Accordingly, provisions were incorporated in the Revised Atomic Energy Act of 1962. While the Health Physics Division of Atomic Energy Establishment, Trombay (AEET, now Bhabha Atomic Research Centre, BARC), could perform a safety review of facilities in DAE, there was practically no expertise outside this organisation for taking up similar reviews in industrial and medical units. To fill this gap, the Directorate of Radiation Protection (DRP) was set up as part of BARC. In 1971 Radiation Protection Rules were formulated for monitoring the use of radiation sources in research, industrial and medical establishments in the country.

As the Tarapur power reactors were nearing completion, a Safety Review Committee (SRC), with specialist members drawn from different divisions of BARC, was set up in 1968 to clear the design and operating procedures for the reactors. In performing this task the SRC was guided by the expertise of USAEC. The SRC carried out the review of Rajasthan power reactors too. By the time IAEA launched its Nuclear Safety Standards Programme in 1974, the SRC gained valuable experience in the field.

In the wake of the reactor accident in Three Miles Island(US) in 1979, the DAE took steps to set up an Atomic Energy Regulatory Board (AERB) as a separate body with responsibility to carry out regulatory and safety functions in an effective manner and with the power to lay down safety standards. The Board, with members drawn from outside DAE also, was constituted in November 1983 under the Atomic Energy Act of 1962. Subsequently, the Board came under the purview of the Environmental Protection Act of 1986 also.

The regulatory and safety systems ensure that equipment at DAE's nuclear facilities is designed according to safety specifications. Even in the unlikely event of any failure or accident, mechanisms like plant and site emergency response plans are in place to ensure that the public is not affected in any serious manner. In addition, detailed plans, which involve the local public authorities, are also in place to respond if the consequences were to spill into the public domain. The emergency response system is also geared to handle independently any other radiation emergency in the public domain.

The members for different committees, such as Nuclear Safety and Radiation Safety and Radiation Protection that assist AERB are drawn from DAE and other sectors. The Health Physics Division of BARC has set up local units in different power plants and other nuclear installations like UCIL for monitoring and providing guidance on radiation protection measures. Several advisory committees are entrusted with specialised tasks like Project Design Safety, Plant Site Evaluation etc. There were several occasions when the AERB took a firm stand on technical matters. Safety of operating plants is vested with the Safety Review Committee for Operating Plants (SARCOP), a DAE unit that interacts closely with the AERB.

The AERB also oversees the safe handling and security of radiation sources in institutions outside DAE, such as hospitals, industries, and research organisations. The Board also provides guidance for the safe disposal of radioactive waste like used radium sources.

In the wake of the recent accident at Fukushima reactors in Japan, the Government of India is planning to create an independent and autonomous Nuclear Regulatory Authority of India that subsumes the AERB. The proposed body will be fully autonomous and will function outside the control of the AEC.

10.8 Health and Safety in Nuclear Power Industry

The nuclear power industry is unique in being the only large-scale energy producing technology which takes responsibility for its wastes. The generation of nuclear energy involves a succession of operations. The main operations are uranium mining, production of Yellow Cake in a mill, conversion and enrichment of uranium to produce fuel in a form suitable for use in the reactors, the operation of reactors for power generation and reprocessing of spent fuel to extract plutonium and unburned uranium and management of the radioactive wastes. This sequence of operations, called the nuclear fuel chain, is not necessarily carried out in a single place, but in different places in a country or even in different countries. In the case of India, for example, uranium mining and milling activities are located in Jharkhand, refining and fuel fabrication in Hyderabad (Andhra Pradesh) reactor operations in several places (Maharashtra, Tamil Nadu, Karnataka, Gujarat, Rajasthan, and Uttar Pradesh) and fuel reprocessing in Tamilnadu and Maharashtra. Japan and some other countries, on the other hand, buy uranium concentrates from Australia or Canada, but get the conversion and enrichment done in the US or Russia. The used fuel from the reactors is sent to UK or France for reprocessing. As a result, the hazardous radioactive cargoes of varying intensity are transported, sometimes transnationally, along highways, sea, and even air routes.

10.9 Chemical Toxicity of Uranium

The chemical toxicity of a substance depends on the way it interacts with the biochemical processes of the human body. The Tolerable Intake (TI) of a substance is defined as 'an estimate of the intake of that substance which can occur over a lifetime without appreciable health risk.' It is usually represented as 'mg per kg body weight.' Some elements like calcium and iron are beneficial and even essential for the body. On the other hand elements like lead, arsenic, and cadmium, are toxic. Since all isotopes of an element behave chemically in the same way, the isotopic composition of an element makes no difference in its biochemical action. Uranium is also injurious to living beings and in this respect, its three common isotopes (234, 235 and 238) are equally bad. Therefore, natural, enriched or depleted uranium makes no difference in the matter of its chemical toxicity. Additionally, the radiations from uranium are toxic.

10.9.1 Inhalation

Pulmonary toxicity of uranium depends upon the chemical form of uranium and the size of the particles. In the case of animals like rats and dogs, the damage to lungs from the inhalation of uranium trioxide is attributed mostly to the radiation than the chemical toxicity. On the other hand, the damage to the respiratory tract on inhalation of uranium hexafluoride vapours is mainly from the hydrogen fluoride (HF) released by hydrolysis of the compound,

$$UF_6 + 2H_2O \rightarrow UO_2F_2 + 4HF$$

In humans, the inhalation of uranium compounds in dust form will cause chemical toxicity depending on their biological solubility. For example, uranium trioxide is more toxic than uranium dioxide. Routine exposure to airborne uranium may not directly lead to increased mortality of human beings. The highly-corrosive HF formed by hydrolysis of uranium hexafluoride inhaled in high concentration causes acute respiratory illness. Increased risk of lung cancer among some uranium miners is more likely due to the element's radioactive decay products radon and its daughters.

10.9.2 Oral Ingestion

Limited information is available on the toxicity of uranium among human beings through oral ingestion. Uranium has a biological half-life (time for half the material to be eliminated from the body) of approximately 15 days. Studies on two groups of persons, one consuming contaminated water containing 0.3–20 and the other 3–500 microgram uranium per day, showed different urinary glucose levels. Since the water also contained other pollutants no definite conclusions could be arrived at.

In recent years, war veterans from the US army exposed to depleted uranium dust in the Gulf War have been found to have elevated urinary uranium levels. Oral ingestion of soluble uranium can lead to renal diseases. Renal effects were occasionally detected among the workers in uranium hexafluoride plants. This problem was, however, not generally reported among uranium miners.

10.9.3 Dermal Absorption

When the body comes directly in contact with a soluble uranium compound like uranyl nitrate, a slow absorption of uranium through the skin occurs. In laboratory experiments with rabbits, death due to renal failure was observed through the dermal application of an ether solution of uranyl nitrate at a concentration level of 28 mg uranium per kg body weight.

10.9.4 Limits for Total Intake (TI) of Uranium

On the basis of studies carried out the values for the tolerable intake of uranium have been prescribed as follows:

Oral: 0.5 microgram per kg body weight per day.

Inhalation: 0.6 microgram per kg body weight per day for soluble uranium compounds and 5 microgram for compounds with limited solubility like UO_3 and U_3O_8.

10.10 Radiation Hazards from Uranium

Uranium itself is not highly radioactive because of its very long half-life. Freshly processed and chemically pure uranium – natural, enriched or depleted – consisting solely of natural isotopes, decays through α-particle emission. These particles cannot penetrate to the sensitive layers of the skin. However, β – and γ – emitting daughter products with shorter half-lives and hence greater activity, start growing from the chemically pure uranium. These radiations not only impact the basal layers of the skin but can penetrate into the body. The uranium in ores, which has been lying in the ground for millions of years is in a state of near equilibrium with a chain of these radioactive daughter products that include radium, polonium, radon etc. Consequently, the radioactivity of mined uranium is approximately 10 times greater than the radioactivity of chemically pure uranium itself. Of all the daughter products of uranium, the α-emitting radon gas with a short half-life of 3.82 days is a major health hazard in uranium mining operations.

10.11 Uranium Mining

Mining uranium is similar to the mining of other metals but, in view of its radioactivity special precautions must be put in place in addition to the usual measures like dust reduction, good ventilation, protection from rock bursts etc.

The post-World War II decade saw a mad rush for uranium, in the wake of the nuclear weapons programmes. Predictions of enormous and cheap electricity from controlled nuclear fission in reactors added to this craze. The uranium rush was considered even bigger than the gold rush. With incentives offered by the US and Canadian governments professional and amateur prospectors literally by the thousands, combed every part of North America (Colorado, Utah, and Great Bear Lake) for uranium deposits. The tropical North Australian territory became the uranium hunting ground for hundreds of adventurers from South Australia. The UK and South Africa also joined the race. The Soviet Union (USSR) commandeered the Czechoslovakian Joachimstahl uranium mines and sent its prisoners of war for digging operations. The Red Army occupied Erzborgue in East Germany and transformed the region into a uranium mining complex, enlisting even criminals to work in the mines.

The hazards of uranium mining and milling were not recognised in the early years. With no precautions put in place, many underground miners became victims of lung cancer through inhalation of the highly toxic radon gas. The widows of these miners even bemoaned that the Geiger counters kept at their homes showed radioactivity in the bodies of the victims.

Empowered by a mandate from the United Nations, the IAEA and ICRP spelled out in 1960, safety measures in mining operations and the limits of radiation to which workers are exposed. Countries engaged in mining operations involving radiation exposure adopted and implemented these measures through their regulatory agencies.

Three types of mining – open-cast, underground and ISL adopted in uranium mining have their characteristic hazards.

Open-cast mining: In open-cast mining, considerable over-burden and waste rock are removed and generally stacked outside the mine. Though low in content the waste rock is not free from uranium. The waste dumps are also sources of radioactive dust. If the dump is in a rain-prone area, the uranium that is leached, especially if it contains sulphide minerals, pollutes the surface and groundwater sources in the area. To avoid this, part of the ore excavated is used to fill back the empty areas in the mine. In some cases, the pits are flooded with water forming a lake. Since the operations are carried out in open pits, the open ventilation helps the dilution of the radon gas, causing minimal harm to the miners.

Underground Mining: In underground mining, access to the uranium-bearing rock is created by digging roadways into the sterile rock at various levels of the mine. Some amount of this waste rock is hauled up and temporarily stacked outside the mine. At a later stage, this waste rock is used to fill the cavities from which the ore is dug out. In underground mines, radon is an important health hazard. The precautions taken to protect the underground miners from radiation hazards are:

- Good forced ventilation to ensure that exposure to radon gas and its radioactive daughter products does not exceed prescribed safety limits,

- Effective control of dust which may contain radioactive constituents,

- Limiting the exposure of workers to radiation levels as low as possible,

- The total effective dose received by a worker should not exceed 20 mSv during a year,

- Installation of radiation monitoring equipment at crucial spots in the mine, and

- Enforcement of strict hygiene standards for the workers, such as wearing dust masks and personal dosimeters (for monitoring radiation exposure).

- In Canada, where some ores contain as high as 5 – 10% uranium, mining activities are carried out mostly through remote operations.

In Situ Leaching (ISL): This involves the extraction of uranium from a porous (sandy) ore material by pumping in a suitable reagent solution and pumping out the uranium-bearing solution. As the amount of solid coming out of the deposit is negligible, the operation causes no major ground disturbance. The miners are practically not exposed to any radiation at the mining stage. However, ISL technology has some hazards. They are,

- Risk of uranium bearing leach solutions escaping beyond the uranium deposit and contaminating water sources,

- Unpredictable effects of the leaching liquid on the host rock of the deposit, and

- The near impossibility of restoring natural conditions after the leaching operations, particularly the quality of water.

Uranium mining operations are by and large conducted following the regulatory norms aimed at protecting the mining personnel. According to the World Nuclear Association, a typical dose rate received by uranium miners in Australia (above background level) is 3–5 mSv per year against the prescribed limit of 20 mSv for radiation workers. In the mines operated by UCIL In India, where the grades of ores processed are lower (0.04% U as against 0.1% U in Australia) the average annual exposure levels reported for 1997 were: Jaduguda – 7.19 mSv, Bhatin – 7.48 mSv and Narwapahar – 6.4 mSv.

Despite these safety measures in uranium mining several citizen groups in many parts of the world, including India, voice their opposition to the uranium mining.

10.12 Milling

The milling operations involve grinding of the ore in a slurry form, selective leaching of uranium and separation of the uranium-bearing solution from the solid waste followed by its concentration and precipitation as the Yellow Cake. All these operations are conducted in a wet form in nearly closed vessels in a more or less continuous stream with the material moving by gravity or by pumping. The small quantity of dust released is trapped in dust filters at exhaust points. The radon emitted during the operations, after dilution with air, is released through chimneys. For this reason, the radiological risks to the operating personnel are less than the risks faced by mining personnel.

Uranium mining and milling units are usually located in areas of low population density. The fine solid residue after the leaching of uranium from the ore in a mill is called 'tailings.' These wet tailings are mixed with treated waste solutions from the mill and made alkaline. The relatively coarser fraction of the solid is then separated and returned to the mine to fill the empty spaces, an operation called 'back-filling.' This fraction may represent 20–40% of the total tailings discharged from the mill. The remaining solid waste product is pumped as a wet slurry to a tail pond. Due to technical limitations, a part of the original uranium (5–10%) remains with the waste. Thorium-230 and Radium-226 are more or less at their original concentration as they are not dissolved in the leaching circuit. The tailings also contain heavy metals and other contaminants such as lead and arsenic as well as reagents like manganese used in the milling operation. Qualitatively all these pose hazards. For example, radon and its radioactive decay products are continuously released. The tailings present in fine particulate form could also find their way into the atmosphere. The Iron pyrite (FeS_2), if present in the ore interacts with water and oxygen, forming sulphuric acid. The acid causes a continuous, slow leaching of metals, including the unleached uranium, which could contaminate the soil and groundwater. However, the degree of hazard depends on the concentration of metals, including uranium in the tailings. Precautionary steps are also taken to minimise these hazards. During the operational life of the mine and the mill, the material in the tailings dam is usually covered with water to reduce surface radioactivity and radon emission. The water layer also prevents the fine dust from being carried by air. To prevent the dispersal of the fine dust into the atmosphere the tailing pond is covered at the end of the mining and milling operations with clay and top soil and vegetation grown thereon. For the low-grade ores like those in India the hazard, however, is not significant.

The tailings contain about 5–10% of unextracted uranium and unleached decay products. Radon gas with a half-life of 3.8 days emanates from uranium. This short-lived hazardous gas is continuously produced from radium-226 which has a long half-life of 1,600 years and hence presents a long-term hazard. Another decay product Th-230 has even a longer half-life of 80,000y. The tailings giving out radioactivity for a long time present a long-term hazard and should be therefore be taken care of for thousands of years. Since the half-life of U-238 is 4.5 billion years, low activity persists for billions of years.

In the early years (the 1950s and 1960s), the mining companies in the US and Canada did not pay attention to clean up the mining and milling sites and tailing dumps after completing the mining operations. These waste dumps came into existence mainly from weapons production activities. When the risks from radiation exposure were driven home by public uproar, the US Congress passed the "Uranium Mill Tailings Radiation Control Act of 1978 (UMTRCA)" for protecting the public from radiation risks through the disposal, long-term stabilisation and control of the mill tailings in a safe and environment-friendly manner. Under this Act, a fund contributed by the federal and state agencies was created for taking steps (including groundwater cleaning) to mitigate the dangers. The US Department of Energy (DOE) cleaned up 24 sites containing waste material piles of volumes ranging from 60,000 to 4.5 million cubic yards.

10.13   Yellow Cake Production

The stages involving the handling of the concentrated solution for precipitation of Yellow Cake, its filtration and drying require strict implementation of the regulatory guidelines such as wearing gloves and dust masks by the operating personnel to avoid the hazards from ingestion and inhalation of uranium.

Yellow Cake, when freshly produced, is safe to handle in a normal way since all the radioactive decay products of uranium are removed at the concentration and purification stages. But in the course of a few weeks, there is a fresh build-up of these products. Precautions are taken for handling this material.

10.14   Refining and Conversion

In a refining and conversion plant, which handles the Yellow Cake for refining and conversion to UO_2 fuel or UF_6 for enrichment, the workers may receive significant levels of internal dose through inhalation and ingestion as they will be dealing with dry uranium compounds in diverse chemical forms. All operations that can cause airborne contamination are carried out in covered equipment with provision for exhaust. It is mandatory for the operating staff to wear respiratory protection. Before the 1970s, the main operations of converting ammonium diuranate to oxide and oxide to fluoride were carried out in fluidised bed reactors. Switching over to rotary kilns for these operations helped in a drastic reduction in exposure doses. In modern plants, the average dose received by a worker is reported to be around 1.0 mSv per year.

As these operations also involve chemical hazards from highly corrosive gases like hydrogen fluoride, fluorine, and uranium hexafluoride, strict guidelines are prescribed for their safe handling.

10.15 Fuel Fabrication

The fuel fabrication operations involve large amounts of radioactive material in diverse chemical and physical forms in conjunction with flammable and reactive chemicals. The major hazard in these facilities is the release of uranium hexafluoride and UO_2. Operating personnel, the public and the environment must be protected from them through adequately designed operating systems. Care is also taken to avoid the fissile material attaining critical size during the fabrication process.

Currently, three types of fuels are used in nuclear power reactors – natural uranium, low enriched uranium (LEU) and mixed oxides of uranium and plutonium (MOX).

Since most of the radioactive daughter products are removed, the refined natural uranium used as fuel in the metal or oxide form is essentially α-active. The precautions to be taken for its fabrication are similar to those followed in refining and conversion plants.

In the low enriched uranium (LEU) fuel fabrication, the radiation from LEU is also low. However, there could be chemical and toxicological effects due to the release of chemicals such as hydrogen fluoride, uranium hexafluoride, hydrogen, nitric acid, and ammonia. Fire and explosions could result in the release of radioactive material.

The MOX (mixed oxides of uranium and plutonium) fuel is a blend of weapons/reactor-grade plutonium and depleted uranium oxide. This fuel is currently being used in 30 reactors (mostly in PWRs) in France, Belgium, Switzerland, and Japan. Several countries – UK, France, Germany, and Belgium (since closed) and India – are also engaged in the fabrication of the MOX fuel. Japan and Russia have plans to commission their plants by 2015. In the US, the construction of a MOX fabrication facility at Savannah River is underway.

The MOX fuel fabrication involves handling large amounts of fissionable (e.g. plutonium) and radioactive materials in dispersible powder form, especially in the early stages. Plutonium isotopes emit neutrons. Strong gamma radiation from Americium-241 (t1/2–432 years), which is formed from the decay of Pu-241 during storage, also poses a hazard.

Fabrication of mixed oxide fuel (MOX) requires special precautions due to the plutonium, whose content can be as high as 15–20%. The MOX dust formed during fabrication is also a matter of special concern. The MOX fuel fabrication plants are provided with at least two physical barriers to prevent the contamination by plutonium. The first barrier is the glove-box and the second the containment building. High-efficiency filters prevent leakage of radioactive particulates into the environment. Precautionary steps are also taken for the safe handling of waste and scrap fuel.

Other major hazards in MOX fuel fabrication facilities are potential criticality, loss of confinement and radiation exposure (both internal and external). The personnel and the public are protected from these hazards through adequately designed safety systems. External hazards include seismic events, explosions and fire, accidental air-crash and extreme weather conditions like flooding.

With the prospects of plutonium-based breeder reactors increasing, greater care is called for in the design and operation of plants handling the MOX and related fuels.

10.16 Nuclear Power Reactors

As of October 2011, there were 432 power reactors with 376,440 MWe total capacity operating the world overclocking 14,655 reactor years of experience and producing about 16% world's electricity needs. Another 63 reactors are under construction. In addition, 152 more reactors are on order or in a planning stage while 350 rectors are on a proposal. (World Nuclear Association, October 2011). This would not have been possible without the extraordinary care and technical expertise that has been bestowed in the design and operation of the reactors. Learning from errors and accidents, the industry has been updating technology and putting in place more and safety procedures to minimise the chances of mishaps.

One of the biggest fears in the minds of the lay public is that a nuclear reactor could explode like an atomic bomb. The fact is that it is impossible for any power reactor to explode because for an uncontrolled explosive chain reaction to occur, the uranium fuel must be highly enriched, much beyond the 4% 235U level used in a commercial light water nuclear reactor. The worst that could ever occur is the core meltdown due to coolant failure. When the coolant system fails the control rods are fully lowered to totally arrest the chain reaction. In the event of the water level descending lower than the fuel level, the heat produced by the fission products in the exposed part of the fuel is enough to drive the core's temperature up leading to its meltdown. The extremely hot (about $2,700^0$ C) molten core could react with water, producing large quantities of steam carrying radioactive fission products. At these high temperatures, the Zircaloy cladding also reacts with water releasing highly inflammable and explosive hydrogen. In the worst case scenario, the hot fuel could even melt the containment vessel and subsequent barriers spewing massive quantities of radioactivity.

Safety systems have been incorporated in reactor designs to meet such contingencies. These include installation of reactor control as an intrinsic feature of the design. The second relates to the efficient extraction of the heat generated by the fission products. More importantly, the reactor unit itself has multiple barriers that contain the radioactive material. As explained earlier, the ceramic uranium oxide fuel pellets themselves act as the first barrier by trapping the radioactive fission products during normal operation of the reactor. The second barrier is the zirconium alloy tube in which the fuel pellets are packed. The third barrier is the thick steel vessel in which the fuel bundles are held. This vessel is housed in a concrete containment structure (building), a gas-tight shell or enclosure around a nuclear reactor to confine fission products that otherwise might be released to the atmosphere in the event of an accident. Such enclosures are usually dome-shaped and made of steel-reinforced concrete of about a metre thick. These structures also protect the reactor against external events and provide radiation shielding when the reactor is in an operational state. Such a steel and concrete structure prevented the spread of radioactivity to the environment in the case of TMI accident though nearly half the core melted. The absence of such containment for the Russian RBMK type Chernobyl Reactor resulted in the release of huge quantities of radioactive material over a wide area. Some reactors are also provided with more systemic features that act as barriers against the leakage of radioactivity in the event of a core melt.

In reactor operations, the safety of the operating personnel is the highest priority. Their exposure to radiation levels is kept to the minimum level through the use of remote handling equipment, and provision of physical shielding. The working times of the personnel working in zones of significant radiation levels are strictly controlled by continuous monitoring of the dose levels.

10.17 Nuclear Reactor Accidents

The accident probability of a power reactor can be kept low by:

- Choice of proper site such that natural calamities like earthquakes and tsunamis and flooding do not cripple the establishment (like in Fukushima),
- Choice reactor types of proven design with diverse fail-safe and back-up safety features,
- Choice of proper construction materials and methods, and
- Selecting highly qualified and trained operating personnel.

Despite all care, it is impossible to design any industrial unit including a nuclear power reactor to have zero accident probability. Nuclear reactor accidents are categorised on an International Nuclear Events Scale (INES) of 1–7, with level 7 rated as the most serious.

Level and Description Effect

1. Anomaly
2. Incident Exposure of a member of the public in excess of 10mSv,
3. Exposure of a worker in excess of the statutory annual limits
4. Serious incident Exposure in excess of ten times the statutory annual limit for workers.
5. Non-lethal deterministic health effects (e.g. burns) from radiation
6. Accident with Minor release of radioactive material, local consequences unlikely to result in implementation of planned countermeasures other than local food controls,
7. At least one death from radiation
8. Accident with wider Limited release of radioactive consequences material likely to require implementation of some planned countermeasures

9. Serious accident Significant release of radioactive material likely to require implementation of countermeasures

10. Major accident Major release of radioactive material with widespread health and environmental effects requiring implementation of of planned and extended countermeasures

Note:

At level 4/5 fuel damage and radiological barrier damage occur.

At level 6/7 significant radiological releases occur.

Source: INES International Nuclear and Radiological Event Scale, User's Manual, IAEA, May 2009.

There have been a number of reactor accidents since the 1950s. The following are some major accidents:

Reactor and Cause Date& INES Scale

- Mayak (USSR) Sept. 29, 1957 (6). Chemical explosion
- Windscale (UK) Oct. 7, 1957 (5) Fire in the graphite reactor
- Three Mile Island (US) March 29, 1979(5) Cooling system malfunction
- Chernobyl (USSR) April 26, 1986 (7) Explosion after an experiment
- Fukushima Daiichi (Japan) March 11, 2011(7). Powerful tsunami following an earthquake of Magnitude 9.0

Mayak Nuclear Complex: A fault in the cooling system of the complex resulted in a chemical explosion and release of an estimated 70–80 tonnes of radioactive material into the air. Thousands of people were exposed to radiation and thousands more were evacuated from their homes.

Windscale Reactor: A fire in graphite core of a reactor resulted in a limited release of radioactivity. The sale of milk from nearby farms was banned for a month. The reactor could not be salvaged and is now buried in concrete. A second reactor on the site was also shut down and the site decontaminated. Subsequently, part of the site is renamed Sellafield where new nuclear reactors were built.

Three Mile Island Nuclear Power Plant: The accident was due to a series of equipment malfunctions compounded by human errors. The problem began with an instrument snag leading to the stoppage of normal flow of feed water to the steam generator. Immediately the backup emergency feedwater system was automatically started. But due to an error on the part of the operator, the water did not reach the steam generator. As a result, there was a temperature and pressure rise in the reactor. But the emergency device provided in the design responded promptly by shutting down the reactor. The cooling process continued with the main coolant pumps continuing to recirculate water in the reactor vessel. The cooling process would have continued without any hitch, but for an error committed by the operator who stopped the main coolant pumps. This was noticed only eight hours later, by which time the fuel heated up resulting in a partial meltdown of the core and generation of hydrogen. Though explosion of hydrogen formed led to the release of a large quantity of radioactive material from the core it was sequestered in the containment building which withstood the explosion. The quantity of the gaseous radioactive material that leaked into the environment from the containment building in the gaseous form was small and hence, did not present a significant hazard to the population exposed. The TMI accident, considered as the worst in the US, led to a major overhaul of precautions regime in the nuclear power industry.

Chernobyl Nuclear Power Plant: The Chernobyl nuclear power plant, which consisted of four reactors of the RBMK-1000 type, is located in Ukraine, 20 km south of the border with Belarus. The accident occurred when the operators ran a test on an electric control system of Reactor No.4. A combination of basic engineering deficiencies in the reactor design and faulty steps by the personnel who were operating the reactor under improper and unstable conditions by switching off the safety systems caused an uncontrollable power surge. This was followed by a series of explosions and fires which blew off the steel container upper plate and caused severe damage to the building. For ten days following the accident the cloud of radioactive dust billowing across northern and Western Europe reached as far as the eastern US. The radiation doses received by the firefighters were estimated to have reached a level of 20,000 mSv, causing 28 deaths within four months of the disaster and 19 subsequently. The UNSCEAR in its report in 2005 on the effects of the accident extrapolated that a total of up to 4,000 people could eventually die of radiation exposure. The agony of the exposed population that went through the ordeal, especially in the absence of accurate information, was pitiful. Though 4,000 cases of thyroid cancer were detected among children, most of them, except nine, recovered through treatment. A team of international experts did not find evidence of any increase in the incidence of leukemia and cancer among affected residents. The UNSCEAR report also says that apart from increased thyroid cancers, "there is no evidence of a major public health impact attributable to radiation exposure 20 years after the accident,"

While the steel and concrete containment structures prevented the spread of radioactivity to the environment in the case of TMI accident though nearly half the core melted, the absence of such containment in the Russian RBMK type Chernobyl Reactor led to the leakage of large amounts of radioactive material.

The damaged Chernobyl-4 reactor is now enclosed in a large concrete sarcophagus. However, the structure is found to be not durable. A New Safe Confinement Structure is due to be installed by 2014. Reactor-2 was shut down after a turbine hall fire in 1991, Reactor-1 at the end of 1997 and Reactor-3 at the end of 2000. While several systemic modifications have been made to improve the safety of all the remaining RBMK reactors, the construction of new reactors of the type has been stalled.

Fukushima Daiichi Power Plant: The plant comprises six separate BWR reactors. On March 11, 2011, an earthquake (9.0 on Richter scale) hit Japan. This was several times more powerful than the worst earthquake the nuclear power plant was designed for. With reactors 4, 5 and 6 already shut down for maintenance, the remaining three units were automatically shut down after the quake and the emergency generators began pumping water into the reactor for cooling. About 45 minutes later, a destructive tsunami with waves 15 metres high (the reactors were designed to handle waves 6 metres high) hit and flooded the entire plant including the emergency generators, the electrical switchgear in the reactor basement and the external pumps in use for pumping the cooling water. The earthquake and the tsunami also hindered all external assistance. With the stoppage of cooling water pumping, the reactor cores were subjected to meltdown, leading to explosions and fires. Despite being shut down at the time of the earthquake, reactors 5 and 6 also began to overheat. Added to this the fuel rods stored in the pools in each reactor building also began to overheat as the water levels in the pools dropped. An estimated 200,000 people have been evacuated from the zone of 20 km radius. With the restoration of the grid power and cooling systems, the workers were able to enter the plant for the first time on May 5. Two workers sustained multiple external injuries and were believed to have died of blood loss. Japan's nuclear workers deserve praise for the dangerous and hard work carried. The exemplary discipline showed by the public in the area while going through the traumatic period during the period also deserves special praise. The accident resulted in the recording of trace amounts of radiation (iodine-131 and Cesium 134/137) around the world.

The task of cleaning up could begin only two years after the accident. Experts working on the site say that it may take 30 to 40 years to just clean up the plant. Decommissioning of the reactors is expected to cost $100 billion while decontaminating the surrounding areas and compensating the evacuated residents would cost up to $600 billion.

The Fukushima accident has prompted widespread public concern about nuclear reactor safety. The IAEA has called upon countries to carry out risk assessment reviews of nuclear power plants through international safety checks to prevent a repeat of such accidents. The US Nuclear Regulatory Commission task force has already recommended several changes in US reactors. These include:

- Requiring that equipment and procedures are in place to keep reactor cores and spent fuel pools cool for at least 72 hours after an emergency and that backup power is available to run cooling systems for at least eight hours if power from outside grid or from emergency generators is lost in a 'station black-out emergency,'

- Requiring that emergency plans address accidents involving multiple reactors on the same site,

- Adding seismically protected systems and instrumentation to assure continued cooling of spent fuels, including at least one source of electric power that can operate cooling pumps and instruments at all times,

- Requiring hardened vent designs for the reactor models (e.g. Fukushima reactors) which could suffer explosions from hydrogen leaks, and

- Strengthening the regulatory oversight of plant safety.
- More regulatory changes will be introduced after further extensive reviews of the accident.

The UN Secretary-General said on April 2011 that the world must prepare for more nuclear accidents on the scale of Chernobyl and Fukushima and that the grim reality will demand sharp improvements in international cooperation and devoting more attention to 'the nexus between natural disasters and nuclear safety.'

10.18 Safety of India's Nuclear Reactors

There have been no major accidents in India's nuclear power reactors. Following the Fukushima accident, the AERB, NPCIL and other bodies have announced that steps have been taken for upgrading the safety of the country's 20 nuclear reactors with three layers of power back-ups, water pipes drawn from off-site locations, elevated water towers and options for injecting nitrogen to prevent explosions. The duration of the power sources/battery operated devices that go into operation in the event of the main power supply failure have been increased. Sensors will be installed in nuclear power plant sites to capture the slightest of seismic activity and automatically shutting down the reactor. The reactors are reported to have passed the structural tests aimed at assessing their capacity to withstand massive seismic shocks. Power plants located on the seacoast will have advance tsunami warning systems. Additional measures include high walls to protect the shores, location of alternate power sources and backup plants at elevated places. With all these safety plans in place, the government considers that the safety of the nuclear plants is assured (Business Standard Feb 2, 2012).

Areva of France, the builders of the 9,900 MWe Jaitapur power plants consisting of four reactors, have agreed to incorporate additional safeguards in the plant in the light of the Fukushima power plant accident experience.

But the related reports are yet to be released for the benefit of scrutiny by the enlightened sections of the society. Meanwhile, public protests against the location of reactors at Kudankulam and Jaitapur are becoming stronger. The Government is sending expert teams to the proposed reactor sites to convince the people of the high standards of safety incorporated or proposed to be incorporated in the reactors at these places.

10.19 Spent Fuel Reprocessing

According to the International Panel on Fissile Materials, ten leading nuclear power producing countries account for 70% of the global stocks of spent fuel (as heavy metal) at the end of 2007 (Table 1).

Of these Canada, Finland, Germany, Sweden, and the US have opted for direct disposal, while France, Japan reprocess part of the spent fuel. The UK also reprocesses its spent fuel, but the future is uncertain. Russia reprocesses part of its spent fuel. South Korea keeps its spent fuel in storage with its disposal policy still undecided.

Table 1: Nuclear Fuel Inventories (end of 2007)

Country	tonnes	Country	tonnes
Canada	38,400	Russia	13,000
Finland	1,600	S. Korea	10,900
France	13,500	Sweden	5,400
Germany	5,850	UK	5,850
Japan	19,000	US	61,000

Source H. Feiveson et al. Bull. Atomic Scientists, June 27, 2011

Spent nuclear fuel reprocessing is an integral part of a closed nuclear fuel cycle to recover uranium and plutonium from the used fuel elements for use as fuel, thereby closing the fuel cycle and gaining about 25% more energy from uranium. Reprocessing is also required to recycle the plutonium and other long-lived actinides in the new generation fast reactors.

The US Department of Energy is developing a new generation of 'proliferation-resistant' reprocessing technology (See chapter 9). Countries having fuel reprocessing facilities include France, Japan, UK, Russia, India, and China. India has developed facilities for reprocessing used fuel from its PHWR reactors for maximising the utilisation of its limited uranium resources. It has also adopted a 'closed circuit fuel management' strategy of plutonium recycling in its fast reactor technology as a part of its three-stage nuclear power programme.

10.19.1 Handling Spent Fuel

The core of a typical 1,000 MWe PWR reactor has about 75 tonnes of low-enriched (3–4%) uranium for its operation. Nearly one-third of this fuel is removed every 12–18 months and fresh fuel is loaded to maintain a steady state operation. The discharged fuel which typically consists of 95% U-238, 1% unfissioned U-235, 1% plutonium and 3% radioactive fission products, is highly radioactive and gives off much heat. After removal from the reactor, the used fuel is stored underwater in specially designed ponds for ten to twenty years. During this period, the relatively short-lived radioactive isotopes decay, bringing down the level of radioactivity and heat generation. The water in the ponds cools the fuel and also provides shielding from radiation. The longer the fuel is stored, the easier it is to handle.

10.19.2 Hazards in Reprocessing

Reprocessing of spent fuel involves cutting the fuel bundles, dissolving the material in acid followed by chemical extraction of uranium and plutonium separately, using suitable solvents, and disposing of the radioactive fission products. While there is nothing extraordinarily difficult in these processing steps, the complications and risks arise out of the necessity of remote handling of all the operations behind heavy shielding due to the high level of associated radioactivity. In the nuclear fuel chain, the reprocessing step is often described as "most risky and dirty."

Compared to reactor operation, maintenance and safety problems are relatively more severe in reprocessing plants. Great attention is required in the maintenance of ventilation systems because the materials to be handled come into direct contact with the ventilation systems. The robustness of the barriers between the radioactive materials and the operating personnel as well as the environment must be ensured stringently. Fire protection and mitigation assume great importance due to the presence of large volumes of organic compounds and combustible gases. Great attention should be paid in handling a variety of processes involving liquids, solutions, mixtures, powders, and gases. Most importantly radioactivity release and the criticality risk of fissionable material call for special attention. Reprocessing technology should also aim at the generation of minimum inventory of wastes amenable for safe interim storage and transport, as well as long-term final disposal.

Through the use of the latest art of technology, it is now possible to design and operate fuel reprocessing plants to achieve the above objectives. Introduction of automated operations has reduced the worker's average annual radiation exposures at reprocessing facilities from over 10mSv to 1.5 mSv per person. The danger of radiation exposure for the public has also been reduced.

10.20 Hazards from Plutonium

Until 1940 no element beyond uranium (transuranics) was known to exist on this planet because all these elements are radioactive with half-lives that are short compared to geologic times. Therefore, these elements which were present when the Earth was originally formed have long since decayed. Plutonium, which is synthetically prepared in nuclear reactors and used in nuclear weapons which are tested extensively in the atmosphere till international agreements came into force is now present everywhere in minute amounts, in the soil, plants, animals and human beings especially in the northern hemisphere.

Plutonium is produced, handled and stock-piled in varying amounts in many countries. Its properties are now understood well. Plutonium separated from used nuclear fuel is a mixture of several isotopes – 239 (major part), 238, 240, 241, 242. All are alpha emitters, except 241, which emits beta particles and changes to americium, an alpha emitter. Plutonium can enter the human body through ingestion, inhalatvion, and injection. Ingestion is not a significant hazard since absorption from the gastrointestinal tract is low. But when it enters the bloodstream through lungs (inhalation) or through wounds, plutonium is preferentially deposited in soft tissues, notably the liver, bone surfaces, bone marrow and other non-calcified areas of the bone. The biological half-life of plutonium is 20 years for liver and 50 years for the skeleton. But plutonium deposited in the gonads is permanently retained, causing genetic damage. The first man-made alpha-active Plutonium-239 with a half-life of 24,200 years is considered to be one of the most hazardous substances. The maximum permissible concentration of plutonium is 4×10^{-12} Curie/cm3 in bones and 2×10^{-11} Curie/cm3 in lungs. A Curie of plutonium weighs 16.4 g.

The total world generation of reactor-grade plutonium in spent fuel is 70 tonnes per year. About 1,300 tonnes of plutonium have been produced in the reactors so far. A major fraction of this remains in the spent nuclear fuel with only 370 tonnes extracted. The dismantling of nuclear weapons, an on-going partial disarmament plan between US and Russia, will give another 150–200 tonnes of weapons-grade plutonium. Currently, 8 to 10 tonnes of plutonium are used each year as MOX fuel by US and European countries as part of a non-proliferation strategy. The fact that several reprocessing plants with high capacities (250 kg Pu per year) are operating in several parts of the world with no major risk should assure many critics that the technology is reasonably reliable.

Contamination, when workers are handling plutonium in any form, is effectively prevented by remote handling in glove boxes, wearing protective clothing and intensive monitoring of working facilities and atmosphere. Currently, the maximum permissible dose is set at 5 micrograms of plutonium for the total body burden. A reprocessing plant, handling fuel from 1,000MWe light water reactor, produces annually about 250 kg plutonium. These figures reflect the magnitude of the problems in the safe handling of this hazardous element.

Keeping the large quantities of plutonium safe is also a formidable task. Fears are expressed that once the fissile material (plutonium or highly enriched uranium) falls into the hands of rogue states or terrorist groups, it is not difficult for them to assemble a nuclear explosive with a blast equivalent to 10–20 kilotons of TNT. Even bombs made from reactor-grade plutonium, though somewhat less efficient, can still be highly destructive. This underscores the need for strict safeguards and physical security over all forms of plutonium (including the spent fuel) and the highly enriched uranium.

In the coming decades, several countries including India and China are taking active steps to promote nuclear energy generation as a viable alternative to fossil fuel energy. A big jump in world nuclear power production could lead to a shortage of uranium resources. In this context, the development of plutonium-based fast breeder reactor technology through reducing costs and upgrading safety aspects is receiving renewed attention. Once this concept gains acceptance, extraction of plutonium and other fissionable actinides from spent fuel elements assumes importance. In such a scenario there is an imperative need for a critical appraisal of fuel reprocessing technology and its upgradation to safe internationally approved standards.

10.21 Accidents at Reprocessing Plants

During reprocessing, the fission products remain in the final aqueous liquid emerging from the complex solvent extraction circuit. This constitutes a high-level hazard. As its radioactivity generates a lot of heat, this waste has to be stored for a long time under conditions of cooling before it can be solidified and readied for disposal. In the past, reprocessing plants around the world have not paid the required attention to safe handling of the wastes from reprocessing operations. In Russia, in 1957, an explosion (level 7) of a waste storage tank in a nuclear reprocessing plant at Mayak (also known as Chelyabinsk Plant) near the city of Kyshtim released 2 million Curies of radioactivity over an area of 15,000 sq. miles, killing 200 people and exposing 500,000 people to dangerous levels of radiation. This incident was made public by the USSR only in 1989! In 1973 in the UK a small explosion in the Windscale (now called Sellafield) reprocessing plant led to substantial spillage of radioactive effluents. In the Hanford military reprocessing plant in the US, the escape of a million gallons of radioactive waste from corroded steel and concrete tanks was noticed only after a long delay. There have also been reports of high levels of occupational exposures in reprocessing plants.

10.22 Radioactive Wastes (Radwastes)

Radioactive wastes (radwastes) may be in the form of spent nuclear fuel itself or the waste separated from reprocessing the spent fuel, as well as radioactive material from institutions such as hospitals, teaching and research facilities, and industrial and commercial units. Radwastes also result from the decommissioning of reactors and other nuclear facilities that are permanently shut down. These wastes are classified into three categories,

- Low level wastes
- Intermediate level wastes.
- High level wastes

10.22.1 Low Level Wastes

The low-level waste comprises rags, tools, clothing, filter etc., which contain small amounts mostly of short-lived radioactivity. It is not dangerous to handle the waste but this must be disposed of more carefully than the normal garbage. Usually, it is burned in shallow sites. When the volume is large, it is compacted before incineration. Worldwide low-level wastes comprise 90% of the volume but only 1% of the radioactivity of all radwaste.

10.22.2 Intermediate Level Wastes

Intermediate level wastes contain higher amounts of radioactivity and may require special shielding. These wastes comprise chemical sludge, waste streams, degreasing, and metal cleaning agents, spent ion-exchange resins, reactor components as well as materials from reactor decommissioning – all contaminated with radioactive fission products – uranium and plutonium. Worldwide these wastes make-up 7% of volume and account for 4% of all radwastes. This waste is first treated, where necessary, to reduce their water content to an optimum level. Certain materials and small items of equipment are compressed while other solid wastes are cut up to reduce their volume. The material is then immobilised in cement based material and packed into stainless steel drums and buried about 20 metres underground. In some situations, the drums are also stored in a suitable geologic formation at a depth of at least 100 metres.

10.22.3 High Level Wastes (HLW)

It is the disposal of high-level radioactive wastes that are of most concern. These high-level wastes are generated at the 'back-end' of nuclear fuel chain and account for 95% of all the radwaste. At the same time, high-level radwaste comprises only 3–4% of the volume of the total radwaste generated in the industry. Albeit its hazardous nature, it is noteworthy that the volume of radwastes from a 1000 MWe nuclear reactor is much smaller when compared with the fly ash and other wastes from a coal-burning 1000 MWe thermal power plant. Following are some typical data:

Nuclear Power Plant

High level wastes: 20–30 tonnes of spent fuel.

The fission products separated after reprocessing and conversion to a vitrified form occupy 3 m^3 volumes

Intermediate level wastes: 310 tonnes

Low level wastes: 460 tonnes

Thermal Power Plant

Fly ash: 3,20,000 tonnes containing 400 tonnes of toxic metals including uranium

Carbon dioxide: 6.5 million tonnes

Sulphur dioxide: 1,10,000 tonnes

Oxides of nitrogen: 22,000 tonnes

The methods for safe disposal of the high level wastes will be described in the next chapter.

10.23 Compensation to Victims of Radiation Exposure

Faced with several demands for compensation to the victims of radiation exposure, the US Congress passed 'The Radiation Exposure Compensation Act 1990 (RECA).' Under this Act, the Federal Government compensates individuals who become sick from accidental exposure to radiation from atomic test fall-out or other sources. RECA recognises five categories of claimants: uranium miners, uranium millers, ore transporters, down-winders, and on-site nuclear test participants. However, the burden of unambiguous proof rests with the claimant.

India's Atomic Energy Act of 1962 does not deal with the liability of accidents involving nuclear power. The Public Liability Insurance Act (1991), which provides immediate relief to persons affected by accidents 'occurring while handling any hazardous substances and for matters connected therewith or incidental thereto' specifically excludes all nuclear – including radiological-accidents. There is only one report on payment of compensation to a person exposed to radiation. An ex-gratia payment of rupees two lakhs was made to the widow of a man who died following exposure to a radioactive material in a scrap yard in the national capital. A framework is therefore needed to deal with domestic as well as cross-border damages in the case of nuclear accidents.

In spite of several inbuilt safety measures in the nuclear reactor systems, there is always a possibility of an incident or accident occurring. Such an unlikely event may result in significant harm to the public and the environment. Promoted by the IAEA, four international agreements have been entered into to provide protection and compensation to the victims of such incidents. These are,

- 1960 Paris Convention
- 1963 Vienna Convention
- 1997 Protocol to amend the Vienna Conventions, and
- 1997 Convention on Supplementary Compensation (CSC)

The Indian nuclear industry which has grown essentially with indigenous efforts (except in the initial stages) under the Atomic Energy Act of 1962 did not sign any of the above conventions. Also, there was no provision in the Act for liability in the event of a nuclear accident. In the wake of the recent international agreements for nuclear power production, the Indian Parliament approved the Nuclear Liability Act in October 2010. Under this Act, a maximum liability was stipulated in the event of a nuclear power reactor accident. Of this, US $10 billion will be on the operator (NPCIL) of the facility and if the cost of the damages exceeds this amount, up to $300 million under Special Drawing Rights will be paid by the Government of India. The operator shall take out insurance for the committed amount before commissioning the facility.

An additional clause in the Act makes not only the nuclear facility operator but also the suppliers of equipment potentially liable for such accidents. This was objected to by US nuclear firms like the General Electric and Westinghouse on the grounds that this clause goes beyond the International Conventions and pushes their liabilities too high. The Russian and French state-controlled companies though having some advantage in this matter over the US suppliers as their governments provide a certain amount of liability protection also expressed some reservations over the clause. India also signed in October 2010, the Convention of Supplementary Compensation for Nuclear Damage, which provides for an international fund to increase the amount available to compensate victims and allows for compensating civil damage occurring within a State's exclusive economic zone, including loss of tourism or fisheries-related income. It also sets the parameters on a nuclear operator's financial liability, time limits governing possible legal action requires that nuclear operators maintain insurance or other financial security measures and provides for a single competent court to hear claims. Russia's concerns over the issues arising out of the nuclear liability law are delaying the deal for the construction of two nuclear power reactors at Kudankulam.

While efforts are on for ironing out the differences to clear the way to the French, Russian and American industries to supply reactors to India, the Indian nuclear suppliers also want the Department of Atomic Energy to waive the nuclear liability law for them. According to them, the equipment machines are manufactured and supplied under close collaboration with and under the surveillance of NPCIL, the operator of the nuclear power plant based on the specifications and quality requirements provided. They further argue that the supplier has no control over preservation, storage, erection and commissioning and operation and maintenance of the equipment supplied. Once the power plant has been built the operation and maintenance is the sole responsibility of the operator NPCIL. The parleys continue!

11

Spent Nuclear Fuel Management and Reactor Decommissioning

11.1 Spent Nuclear Fuel Management – An Overview

Spent nuclear fuel management comprises an integrated series of technical operations, starting with the discharge of the spent fuel assemblies from a power reactor and ending with direct disposal (open or once-through fuel cycle) or with reprocessing (closed fuel cycle). Direct disposal (once-through fuel cycle) involves placing the spent fuel in a location such as a geologic repository for periods extending up to 200,000 years under conditions which would not permit its removal. Another variant is interim storage with continuous monitoring enabling its retrieval at a later date for either direct disposal or reprocessing.

In a closed nuclear fuel cycle, the spent fuel is reprocessed and the uranium and plutonium recovered are either be added to the front-end of the fuel cycles or used to produce mixed oxide (MOX) fuel. Currently, some MOX fuel is being used in LWR reactors. Ultimately, it will be used in new-generation fast reactors. The high-level waste (HLW) left behind contains fission products and minor actinides. After immobilisation in a glass matrix, it is safely stored until its radiotoxicity will have decreased to safe levels.

The US was the first country to set up nuclear reactors, reprocess the irradiated fuel from the reactors and recover plutonium for assembling atom bombs. Russia, France, and the UK followed suit. With the anticipated growth of breeder reactor technology involving the recycling of separated plutonium and uranium, the closed fuel cycle concept held out. But with delays and cancellation of the breeder programmes, the use of plutonium (e.g. Pu as MOX) was limited to the thermal reactors already operating in Belgium, France, Japan, Russia, Ukraine, Bulgaria, China, Germany, Hungary, India, Italy, Netherlands, UK, and Switzerland. The recovered uranium is recycled by Russia and some countries.

The countries that opted for direct disposal of the spent fuel (once-through fuel cycle) include Canada, Finland Czech Republic, Germany, Lithuania, Romania, Russia, South Africa, Spain, Sweden, Ukraine, and the US.

With most of the spent fuel pools full or nearly full, and delays in reprocessing and identification of safe repositories, another strategy that has been adopted is dry cask storage for spent fuel. The fuel assemblies are placed in steel canisters that are surrounded by a heavy shell of reinforced concrete with vents that allow cooling air to flow through to the wall of the canister. These casks are kept in buildings for additional protection against weather damage, accidents, and attack. According to IAEA "the long-term (dry cask storage) (is) becoming a progressive reality…storage up to 100 years and even beyond is possible." Some countries that have adopted this strategy include Argentina, Belgium, Brazil, Bulgaria, Czech Republic, Hungary, Italy, South Korea, Mexico, Pakistan, Slovenia, Switzerland, Ukraine, and the US.

11.2 Worldwide Spent Nuclear Fuel Inventory

When taken out of a reactor, a fuel bundle will be emitting radiation, principally from fission fragments, and heat. It is unloaded into a storage pond immediately adjacent to the reactor to allow the radiation levels to decrease. This process is called 'cooling.' The water in the ponds shields the radiation and absorbs the heat. Between four days and one year after discharge, the heat output of a used LEU fuel decreases roughly by a factor of ten. Ten years after discharge, it is down by a further factor of ten. A hundred years after discharge, it is down by another factor of five. For much of the first 100 years, the radioactivity is dominated by the fission products, predominantly strontium-90 and caesium-137 with half-lives of around 30 years. After a few hundred years, the gross activity is dominated by the transuranics: plutonium, americium, neptunium, and curium.

Currently, about 10,500 tonnes of heavy metal (tHM) (spent nuclear fuel) are unloaded every year from nuclear power reactors worldwide. This discharge rate is estimated to have risen to about 11,500 tHM by 2010. The term tHM indicates the fuel mass as measured by its original uranium and plutonium content and does not include the weight of structural materials or the oxygen in the uranium and plutonium oxides. The total amount of spent fuel cumulatively generated worldwide by the beginning of 2004 was close to 268,000 tHM of which only 90,000 tHM has been reprocessed. By 2010, the cumulative figure is projected to reach close to 340,000 tHM with a corresponding increase in reprocessed fuel. The US with 64,500 tonnes including 15,335 tonnes stored in dry casks, has the biggest stockpile of spent fuel from reactors. More than 90% of spent fuel in the world today is stored in pools at reactor sites or away from reactor facilities. By the year 2020, when most of the presently operating nuclear power reactors will approach the end of their permitted lifetime, the total quantity of spent fuel generated will be of the order of 445,000 tHM.

Current commercial nuclear fuel reprocessing capacity is shown in Table-1. The annual capacity, which now stands at 5,630 tonnes is projected to rise in future to 6,525 tonnes with China also developing these facilities.

Part of plutonium recovered from the used fuel as well as the weapons-grade plutonium from military sources is being recycled into MOX fuel. The inventory of recyclable materials is shown in Table 2.

Table 1: World Commercial Spent Fuel Reprocessing Capacity (tonnes/y)

LWR Fuel France (la Hague)	1,700
UK (Sellafield)	900
Russia (Ozerisk)	400
Japan (Rokkasho)*	800
Other Fuel UK, Sellafield (MAGNOX)	1,500
India (PHWR)	330
Total	**5,630**

* Expected in October 2012

Source: World Nuclear Association. May 2012.

Table 2: Inventory of separated recyclable Materials Worldwide (tonnes)

Description	Quantity	Natural Uranium Equivalent
Plutonium from reprocessed fuel	320	60,000
Uranium from reprocessed fuel	45,000	59,000
Ex-Military plutonium	70	15,000
Ex-Military HEU	230	70,000

Source: World Nuclear Association, May 2012.

11.3 Once-through vs. Closed Fuel Cycle

There has been a keen debate about the best approach to manage the spent fuel from nuclear power reactors, whether to dispose of it directly in safe geologic depositories (once-through fuel cycle) or reprocess it to recover and recycle the plutonium and uranium, disposing of only the wastes from the reprocessing and recycling (closed fuel cycle).

Preference to closed fuel cycle that enables recycling plutonium and unspent uranium is based on the following considerations,

- The world's present viable uranium resources are of the order of 5.4 million tonnes. The current demand for uranium is about 68,000 tonnes per year. With conventional reactors now in use, the uranium reserves are enough for about 80 years. With renewed interest in nuclear power in some countries, the demand for uranium is bound to rise further. Reprocessing the spent fuel for recovery of plutonium and uranium and recycling for power generation could result in a 25% saving of natural uranium resources.

- Recycling is claimed to result in a five-fold decrease in the volume and ten-fold decrease in the toxicity of the wastes that have to be sent to geologic depositories for safe storage.

- Reprocessing is necessary for the development of new generations of reactors (some of them are Fast Breeder Reactors), in which plutonium and other actinides are used for energy generation, thus increasing the energy output for a given amount of nuclear fuel by a large factor. It will help reduce consumption of natural uranium and also save the energy necessary for the enrichment of uranium (LEU) for use as reactor fuel.

The arguments advanced against reprocessing are,

- The complex and expensive facilities to extract the plutonium make reprocessing economically unviable in view of the current availability of uranium costing more than a once-through fuel cycle and will do little to reduce the amount of long-lived radionuclides in the waste.

- A July 2000 report commissioned by the French Government concluded that reprocessing is uneconomical, (Annie Makhijani; Science for Democratic Action, Vol.9, no.2). An MIT Report published in 2003 also says that the disposal cost of a once-through fuel cycle is four times less than the cost of reprocessing. According to one estimate, the current price of US $1,000 per kilogram of heavy metal reprocessing and recycling in existing LWRs will be more expensive than the cost of direct disposal of spent fuel until the uranium price exceeds $360 per kilogram, a price that is not likely to be seen for a few decades. Even in the wake of the recent emergence of buying interest and upward momentum of demand, the spot uranium price during the first week of September 2011 stood around $115.5 per kg.

- While spent fuel emits lethal radiation, separated plutonium can be handled easily making it possible for the terrorists to steal it and assemble a bomb.

- Reprocessing does not eliminate the requirement for a geologic repository or even reduce its size significantly. This is because, in effect, reprocessing involves managing not only spent fuel but also the high-level waste generated in reprocessing. Currently the plutonium recovered from thermal reactors and dismantled nuclear weapons are being recycled in the form of blended MOX. Because of the high radiation and heat content of the used MOX fuel, as well as the accumulation of undesirable uranium and plutonium isotopes, only one round of recycling of this MOX is feasible. Ultimately this spent MOX fuel as well as the fission products and the residual uranium separated from the original spent fuel must be safely disposed of.

The choice of spent fuel strategy is thus a complex decision involving many factors – political, economic, safeguards and environmental protection.

11.4 Nuclear Waste Management during the Early Phase

During the early phase of nuclear weapons production, the US and other countries followed simple methods like shallow pit burial of the waste material without much concern for the adverse effects. For example, the Hanford unit in the US, built over an area of 560 sq. miles to produce plutonium for nuclear weapons discharged its wastes in its vicinity for decades. As a result, Hanford's Nuclear Reservation Zone has become America's most polluted site containing huge quantities of radioactive wastes. The US EPA has recently drawn up a 30-year time-table for cleaning up the area at an estimated cost of $50-100 billion. Some waste was also dumped into the sea with minimum treatment like dilution. This practice was later banned through a moratorium accepted at the London Dumping Convention, 1983. The IAEA laid down mandatory guidelines for dealing with radioactive wastes for implementation by all nations. These include:

- Securing acceptable levels of protection of human health and the environment with negligible effects beyond
- national boundaries, and
- Minimal impact on future generations.

11.5 Current Status of Spent Nuclear Fuel and High Level Waste Management

At present, the most satisfactory method for disposing of the spent fuel and other high-level wastes is considered to be storage in geologic repositories. The used fuel is stored in cooling ponds underwater for as long as possible, say 25–50 years, to allow most of the short-life nuclides to disintegrate. Pending the identification of suitable geological repositories, the material is kept in dry storage facilities. The geologic repositories chosen are considered adequately passive towards the material interred and require no continued human involvement for their safety. Since the unprocessed fuel contains, in addition to fission products, plutonium-239 ($t_{1/2}$: 24,200 years) and plutonium-240 ($t_{1/2}$: 6,800 years) and other long-lived transuranics, the repository should be able to contain the waste for a very very long time, say 200,000 years.

For the processed high-level waste (HLW) the containment requirement could be about 500 years, at the end of which period most of the radioactivity will have decayed leaving behind an activity similar to that of a corresponding amount of naturally-occurring uranium ore from which it originated. This HLW, which is much smaller in volume, is vitrified in borosilicate glass or ceramic material, and suitably encapsulated in stainless steel canisters. After covering with bentonite clay, the canisters are lowered to depths of 500 metres or more into deep geologic repositories that have retained their integrity for millions of years.

Some geological properties that make a particular formation a good choice as a repository for spent fuel and HLW are:

- The geological formation should be water-tight for at least a million years.

- In the event of leakage, the migration of long-lived activity must be negligible.

- Information on the effect of high radiation levels on the physical properties of the containment material must be available.

- The effect of heat generated by radiation on the bearing capacity of the containment material must be negligible.

- The solubility of the glass or ceramic material in fresh water, or salt solutions moving through the burial area in future must be negligible, and

- The spent fuel must be amenable for retrieval at a future date when reprocessing may become a viable proposition or when the performance of the repository raises doubts.

The geological formations considered suitable for the purpose are,

- Evaporative beds like rock salt formations,

- Fine grained sedimentary rocks like shale which have

- good plasticity, low permeability and good

- ion-exchange characteristics, and

- Igneous metamorphic rocks such as granite and basalt.

Many countries have invested significant time, funds and energy in identifying suitable locations as depositories. To begin with, a multinational geologic repository was mooted in the 1990s. Two sites – Marshall Islands and Palmyra Island in the Pacific and, a third site in Western Australia were identified as depositories. But these proposals were dropped in the face of determined public opposition. With no country willing to host a multinational spent fuel repository, individual countries directed their efforts to identify suitable locations in their jurisdiction. But only a few have been able to take firm decisions. The present status in respect of some countries is summarised below.

US: In 1977 the US, under a directive from President Jimmy Carter stopped reprocessing spent nuclear fuel and adopted strictly the once-through fuel cycle strategy as part of its non-proliferation policy. In 1981, President Reagan lifted the ban on commercial reprocessing. In 2001, President Bush permitted the development of "reprocessing and fuel treatment technologies that are cleaner, more efficient, less waste intensive and more proliferation resistant."

The US has also been pursuing the once-through fuel cycle strategy. The spent fuel from the reactors is kept in unprocessed state at the reactor sites in cooling ponds or in casks kept in concrete silos. After extensive studies, a repository was selected at Yucca Valley Mountain in Nevada. This is a particularly dry area in which the casks carrying the waste are to be located nearly 500 metres deep. Water seepage is considered minimum and even if the metal casks were to corrode. The scanty moisture at that depth prevents migration of the contents to the surface. The site chosen is also far away from the populated city of Las Vegas. But the Nevada civil society voiced strong opposition to the location of the repository in the state, citing the risk of the nearby dormant volcanoes turning active in a few thousand years. Fear was also expressed the multiple barriers might breach resulting in a leak of radioactivity during the long internment period. Deterred by the opposition from civil society the state and federal governments put the proposed project on hold.

UK: There was no serious planning for the long-term management of waste generated by military and civil nuclear programmes till 1976 when a Royal Commission recommended the formation of a Nuclear Waste Disposal Corporation for the purpose. A few sites identified by the Institute of Geological Sciences could not be investigated in the face of strong public opposition. Dumping of low and intermediate-level waste even in deep mines was abandoned in the face of strong public protest. Finding no way out, the UK Government decided in the 1980s that the vitrified HLW sealed in stainless steel canisters would be stored for at least 50 years by which time the radioactivity would have decayed to some extent making it suitable to safely dispose of the material. A new Committee for Radiation Waste Management was constituted in 2002 to review the proposals in an 'open, transparent and inclusive manner' and to engage the public in a discussion to find an acceptable method for the safe disposal of the material.

France: The spent fuel is reprocessed and the plutonium is used in LWR reactors in the form of MOX fuel. The spent MOX fuel is being stored pending the development and commercialisation of new generation fast reactors. For safe storage of HLW and other long-lived wastes from reprocessing, a 500-metre deep geologic repository in the clay formations at Bure in Eastern France is being planned. This site is expected to start functioning by 2025.

France and UK undertake spent fuel reprocessing for other countries. These countries now stipulate that the high-level waste from reprocessing and the recovered plutonium and uranium must be taken back by the country of origin.

Russia: Russia's spent fuel management policy is based on the closed nuclear policy. About 50 tonnes of recovered reactor plutonium and 34 tonnes weapons plutonium are being stored for use in the new generation fast reactors. Russia reprocesses its spent fuel and also the spent fuel from power reactors provided by Russia. The waste from reprocessing and plutonium can be left in Russia. There are two large spent storage pools in Zheleznogorsk, near Krasnoyarsk in Siberia. Russia is also building a central dry cask storage capacity there. There is no active programme to site or build a geological repository for either spent fuel or HLW though some repository sites are reported to be under examination.

Japan: Spent fuel reprocessing policy is adopted by Japan. The spent fuel and HLW are kept at the storage facility at Rokkasho. According to one report, more than half of the storage pools for spent fuel at nuclear power plants throughout Japan will be at full capacity if the reactors continue to operate. A reprocessing plant with a capacity of 800 tonnes per year has also been built here but it has yet become fully operational. Another storage facility coming up at Mutsu has been put on hold in the wake of the accident at Fukushima. Meanwhile, problems are being faced in handling the storage pools due to failures in the cooling systems. A retrievable storage facility is slated to be operational by 2035.

Canada: No site has been identified for long-term storage. Meanwhile, all spent fuel is stored at the reactor pools and in canisters.

Sweden and Finland: Both these countries, which earlier initiated proposals for outsourcing reprocessing, have now opted for the once-through cycle. They now plan to use copper casks embedded in bentonite clay and store them in geologic repositories. Finland and Sweden have identified 1.9 billion-old granite repositories for the purpose. Finland has identified a repository near Olkiluoto, which will be opened in 2020. Sweden has chosen a repository near Forsmark.

Germany: Until 2005, Germany has been getting its spent fuel reprocessed in France and UK. The HLW from reprocessing is stored at the interim storage facility at Gorleben. After the Fukushima accident, Germany has decided to shut down all its reactor operations by 2022.

A matter of concern is that most of the spent fuel pools in the US and other countries are becoming full or nearly full. An accident involving spent nuclear fuel can be as disastrous as that caused by reactor core meltdown. The consequences can be even more disastrous if an accident occurs in the pools packed with recently discharged spent fuel. The damage to the spent fuel pools in the recent accident at Fukushima highlights such a risk. With huge leaks in the pools in which the spent fuel is kept, it is becoming difficult even two years later to keep the Fukushima fuel pools in a safe state. Huge quantities of cooling water contaminated with radiation are also leaking into the pits and into the sea. It is noteworthy that while the six reactors at Fukushima had 60 to 63 tonnes each of spent fuel on site, a single unit, Vermont Yankee, located in Vermont, US has a staggering 690 tonnes of spent fuel on site!

India: With the adoption of the 'closed nuclear fuel cycle' strategy involving regular reprocessing, India recovers the plutonium from the spent fuel of its reactors. Waste management facilities have been set up at all reactor sites – Trombay, Tarapur, Kalpakkam, Narora, Rawatbhata, Kakrapar, and Kaiga. The used fuel is stored underwater in ponds at reactor centres. Averaging over the past 20 years the share of nuclear energy in India has not exceeded 2,000 MWe per year. Therefore the quantity of waste generated so far is very small compared to other countries. The low-level liquid waste is retained as sludge after chemical treatment. Some of the low and medium level waste is fixed in cement matrix and buried in engineered trenches at shallow depths. The reinforced concrete trenches are waterproofed and covered with reinforced concrete and kept under continuous surveillance. The HLW solutions are either held as such or evaporated and vitrified into glass and sealed in metal blocks for storage. From all available accounts, a disposal site and methods are yet to be finalised. As the generation of nuclear energy is expected to increase substantially (more than 20,000MWe by 2020), in the coming decades, a safe and permanent HLW disposal strategy must be put in place soon.

11.6 Nuclear Reactor Decommissioning

All power plants including nuclear power plant have a finite life beyond which it is not economically feasible to operate them. The average life of 400+ power reactors that are currently operating worldwide is put at 25–30 years. In Western Europe 75% of the plants are in the last half of their operating life. In the US, 90% of the power plants are already more than 20 years old and half of these have been operating for more than 30 years. The International Atomic Energy Agency puts the average lifespan of a nuclear power reactor at 40–50 years. This means that already majority of the power reactors are reaching the end of their operating life. With proper management, vigilance, and safety enhancement nuclear power plants can perhaps operate beyond this period. Some countries are already taking steps in this direction. Nuclear power plants are mandated to renew their exploitation license every ten years. But all reactors will come to the end of their lifespan when they should be shut down and steps taken for their decommissioning.

Considerable experience has been gained over the last 20 years in the decommissioning of 102 nuclear reactors representing different reactor technologies and ranging in size from small prototypes to large commercial plants. These include 11 units that suffered accidents to a point where their repair is not economically justified and 25 units closed prematurely on political grounds. The IAEA reports that at the end of 2005, eight power plants have been completely decommissioned and dismantled and the sites released for unrestricted use. Additionally 17 power plants have been partly decommissioned and dismantled with the sites to be released eventually. Thirty plants undergoing minimum dismantling will be subjected to long term enclosure.

The objective of decommissioning is to restore the nuclear facility to a condition where there is no risk to the public health and safety or the environment. The process of decommissioning may vary from site to site. It involves removing the spent fuel, dismantling any systems or components containing radioactive products, and cleaning up or dismantling contaminated materials. All activated and contaminated materials have to be removed and sent to a waste processing, storage or disposal facility. If a major decommissioning activity does not meet the prescribed regulations, the activity is stopped till the stipulated conditions are fulfilled.

The IAEA has spelled out three options for decommissioning of nuclear power plants. These are:

Immediate Dismantling (Early Site Release/Decon): This allows for the facility to be removed from regulatory control soon after shutdown. This will be followed by final dismantling and decontamination activities in few months or years. The site is then released for reuse after clearance from the regulatory authority.

Safe Enclosure (Safestor): Under this option the regulatory control is in force for 40 to 60 years, with the facility placed in safe storage configuration till dismantling and decontamination.

Entombment (Entomb): This involves placing the facility into a condition that will allow the remaining on-site radioactive material to stay put without the requirement of permanent total removal. After reducing the size of the area where the radioactive material is located, the facility is entombed in a long-life structure such as concrete that will last for a period sufficiently long for the radioactivity to decay to safe and permissible levels. Entombment involves long-term maintenance and monitoring.

The entombment strategy for a nuclear power plant is rarely applied to a nuclear power plant, because the size of a nuclear power plant is too large to be simply entombed. Entombment is however adopted for small reactors and smaller nuclear facilities. The Chernobyl (Chernobyl) nuclear reactor which suffered a serious accident leading to the core meltdown had to be entombed because there was no other choice. But 25 year later the concrete containment (called *sarcophagus*) is reported to be crumbling. Fears are expressed that it could even collapse, releasing another radioactive cloud into the air. Examples of some small reactor units that were subjected entombment are,

- Lucens, Switzerland.
- Dodewaard, Netherlands
- Boiling Nuclear Superheater Facility, Rincón, Puerto Rico
- Hallam Nuclear Power Facility, Hallam, NE, US
- Piqua Nuclear Generating Station, Piqua, OH US

11.7 Meeting the Costs of Decommissioning

The costs of decommissioning nuclear power reactors are very high and vary greatly, depending on the reactor type and size, its location, the proximity and availability of waste disposal facilities, the intended future use of the site, and the condition of both the reactor and the site at the time of decommissioning. Substantial costs are incurred in the removal, treatment, and disposal of the major components of the contaminated unit such as valves, pumps, piping, steam generators, and the reactor vessel. Decontamination of floors, walls, and equipment also add to the costs. The occupational dose received by the personnel during the exercise should also be considered as a cost.

In the US, the dismantling costs generally range from $300 million to $400 million per power reactor. If the operation is carried over a long period extending up to 60 years, the costs can go up to even $1 billion. In France, decommissioning of the 70 MW Brennilis power plant has already cost €480 million and is yet to be completed. In the UK, decommissioning of the 32 MW Windscale Advanced Gas Cooled Prototype Reactor cost €117 million.

The US Nuclear Regulatory Commission requires the plant owners to set aside money to pay for the future decommissioning costs when the plant is still operating. The licensees may demonstrate financial assurance for the purpose by adopting one or more of the following measures:

Prepaymet: A deposit at the start of operation in a separate account,

Surety, insurance or guarantee: Assurance that the cost of decommissioning will be paid by another party, should the licensee default,

External sinking fund: A separate account (called the Nuclear Decommissioning Trust Fund) outside the licensee's control to accumulate decommissioning funds over time, if the licensee recovers the cost of decommissioning through rate=making regulations (establishing rates of payment for the utility) or non-bypassable charges (Charges to the consumers receiving service from the utility). The purpose of the Trust Fund is to spread the high cost of decommissioning to the ratepayers over the life of the plant by taking advantage of compound interest on the funds invested in deep discount bonds.

There is still no legislative framework in India with regard to the funding of nuclear power plant decommissioning activity (CAG Report No. 9 of 2012–13). The Department of Atomic Energy has issued a notification in December 1988 to levy a decommissioning charge of 1.25 paise/kWh energy sold. In October 1991, the levy has been revised to 2 paise/kWh. No rules were, however, framed providing for the creation of Decommissioning Trust Funds by the utilities.

With several nuclear reactors coming to the end of their life and decommissioning becoming an attractive business, several companies specialising in the area have entered the field.

12

Nuclear Power and Public Perception

The transformation of uranium from glass and ceramic colouring material to a viable energy resource for power generation has been described in these pages. According to World Energy Outlook 2012 of the International Energy Agency, electricity demand is increasing twice as fast as overall energy use. Electrical power generation grew at an average rate of 3.4% from 1073 to 2010 and is projected to grow by 87% from 2010 to 2035. The increased demand is most dramatic in the developing economies in Asia (India and China), projected to grow at an average annual rate of 4.3%. Nuclear energy is considered the most environmentally benign way of producing electrical power on a large scale. But the public continues to wary of the safety of nuclear power.

Public opinion on nuclear power has been a complex and controversial subject. It has gone through marked changes ever since it has come into being. But the indisputable fact is that the gamut of issues is always looked in the backdrop of the destruction of the cities of Hiroshima and Nagasaki by the atom bombs. On the one hand, nuclear energy has been championed as the saviour of the world from energy famine while on the other hand, it is considered as the cause for the civilisation to vanish in a cataclysmic holocaust. For a long time, nuclear power reactors have been claimed by their proponents to be free from accidents. But in the wake of the accidents at the Three Mile Island, Chernobyl and more recently at Fukushima the safety of nuclear power plants has become an emotional issue in the energy debate.

Global energy use has been climbing steadily over the years with expanding industrial economies. The current per capita energy consumption is about 100 times higher than it was 2,000 years ago. This rapid rise is expected to continue for several years. It is generally agreed that the welfare of people, those living now and the generations to follow, is dependent on energy security. The present global electricity generation is 20,180 TWh (terawatt hours). It is projected to increase to 34,291 TWh in 2030. Most of the projected growth will come from developing countries (IAEA). For example, India's primary energy consumption more than doubled between 1990 and 2011. Despite this, the country's per capita energy consumption remains much lower than that of many developed countries. In the context of global warming concerns and rise in oil prices over the last decade the best choice for India to meet its energy needs in the coming decades is through having an energy mix in which nuclear power is an important component. In the words of Prime Minister Manmohan Singh:

"For a large and fast growing economy like ours, and given the volatility and uncertainty of international energy markets, it is in our national interest that we tap all sources of energy and diversify our energy mix. Nuclear energy is one option to enhance our energy security." (Times of India, October 8, 2011).

The acceptance of the nuclear power industry hinges on its safety standards because an accident anywhere in the world would stoke an anti-nuclear backlash amongst the public everywhere else. Other factors that have an adverse impact are the by-products of nuclear energy – radioactive waste, nuclear terrorism and non-proliferation. It is therefore essential that the highest priority is given to the development of nuclear reactors of safe design, quality workmanship, and diligent operation. To quote Elena Sokova, one of the Directors of the James Martin Center for Nonproliferation Studies, *"The nuclear establishments must teach the fundamentals of security and non-proliferation to every future specialist in every country that possesses a civilian nuclear research programme."* (Bulletin of Atomic Scientists, April 1, 2010). Steps should be taken to solve the challenge of the safe disposal of the huge amounts of radioactive wastes. It is also necessary to constitute a truly independent authority to review the safety standards of a nuclear plant. Further, in an emergency, each reactor should be able to stand on its own through avoidance of common facilities.

The subject of nuclear power, due to its highly technical nature, is normally outside the realm of the common man. India is experiencing, for the first time, a demonstration of how easy it is for the anti-nuclear activists to rouse public opinion by playing on the mysteries of the atom and the dangers of nuclear radiation. The public reaction towards nuclear power is currently based on what the community learns through media and the environmentalists. For this reason, it is imperative that correct information is conveyed to the public through improved communication between the government, nuclear industry, scientific and technological community, mass media, environmentalists and political forces in a transparent manner. The nuclear scientists should bear "the burden of translating their knowledge into a persuasive public explanation." This kind of education should be broad-based with extra attention paid to the politicians and the media. A major complaint is that the nuclear establishment in India, as elsewhere, has been extremely secretive and tight-lipped about its workings. It has even been described as "a state within the state." The civil nuclear power authority should make every effort to erase this impression from the minds of the public through dissemination of relevant factual information to the public in a transparent manner. The Chairman of Indian Atomic Energy Commission recently admitted that there were shortcomings in the nuclear establishment's efforts to reach out to the locals of Kudankulam who demonstrated against the nuclear power plant (Times of India, Oct.18, 2011). Scientists and scientific organisations and societies should also help government, media, students and the lay public in grasping nuclear issues and confidence through extensive exhibitions, workshops, seminars and the state of the art audio and visual aids. The Nuclear Power Corporation is a commercial enterprise entrusted with the task of supplying nuclear power to the public. It has been even raising funds by borrowing from the public for augmenting its working capital. In this age of 'transparency revolution,' secrecy surrounding the civilian nuclear power and preventing scrutiny by the public, the major stakeholder, has no relevance. Continued public opposition could even lead to situations where even near-complete reactors are halted creating energy crunch and economic slowdown. For instance, even the all-powerful Chinese administration had to yield to public protests and cancel the $6 billion project in Longwan Industrial Park designed to have a manufacturing capacity of 1000 tonnes of uranium fuel for supply to nearly half of China's nuclear power plants (Times of India, July 14, 2013).

In the absence of proactive steps, fears and misconceptions about nuclear power among the public at large will continue to be a major hurdle for the growth of the civilian nuclear power industry. Yukiya Amano, Director General of IAEA observed,

"We need to share both good news and bad news to get a better understanding of the problem. We have to explain complicated things in simple manner. It is only through better communication and higher transparency that we can achieve this." (Times of India, Dec.31, 2012).

Additional Reading

F. Soddy, *The Interpretation of Radium*, John Murray, London (1912).

H.G. Wells *The World Set Free*, Macmillan, London (1914).

S. Glasstone *Source Book on Atomic Energy*, Affiliated East-West Press, New Delhi (1967).

Norman Moss *The Politics of Uranium*, Universe Publishers, New York (1984).

Tom Zoellner *Uranium, War, Energy and the Rock that shaped the World*, Viking, London (2009).

Ross Annabel *The Uranium Hunters*, Rigby, Adelaide (1971).

Richard Rhodes *The Making of the Atom Bomb*, Simon & Schuster, New York (1986).

J. Bernstein Plutonium, *A History of the World's most dangerous Element*, Joseph Henry Press, Washington DC. (2007).

P.D. Wilson (Ed), *The Nuclear Fuel Cycle, From Ore to Waste*, Oxford University Press, Oxford (1996).

IAEA, *Uranium Extraction Technology* Technical Report Series 359, IAEA, Vienna (1997).

G.F. Knoll *Radiation Detection and Measurement*, Wiley, New York (1979).

S.S. Kapoor and *Nuclear Radiation Detectors*, V.S. Ramamoorthy New Age International, New Delhi (1993).

S. Villani (Ed), *Uranium Enrichment*, Springer Verlag, Berlin (1979).

S. Glasstone & *Nuclear Reactor Engineering*, A. Sesonske van Nostrand, New York (1981).

John R. Lamarsh & *Introduction to Nuclear Engineering.* Anthony Baratta Prentice-Hall, New York (2001).

J. Bernstein, *Nuclear Chemical Engineering.* Thomas Pigford & McGraw-Hill, New York (1981).

H.W. Levi W. Marshall *Nuclear Power Technology (3 vols.)* Clarendon Press, Oxford (1983).

D.N. Lepedes McGraw-Hill Encyclopedia of Energy, McGraw-Hill, New York (1971).

B.R.T. Frost *Nuclear Fuel Elements*, Pergamon Press, Oxford (1982).

Additional Reading

IAEA *Nuclear Power Reactors in the World*, IAEA, Vienna (1978).

Mohamed Mohamed *Nuclear Energy Conversion*, El-Wakil American Nuclear Society (1978).

Joel Weisman *Elements of Nuclear Reactor Design*, Elsevier, Amsterdam (1977).

James A. Lake, *Next Generation of Nuclear Power*, R.G. Bennet & Scientific American, Jan 26, 2009.

John F. Kotek Stephen Goldberg & *Nuclear Reactors, Generation to* Robert Rosner *Generation*, American Academy of Arts and Sciences, Cambridge, MA (2011).

H.N. Sethna *India's Atomic Energy Programme- Past and Future*, IAEA Bulletin, 21(5) (1979).

R. Ramanna *Years of Pilgrimage*, Viking, New Delhi (1991).

M.R. Srinivasan *From Fission to Fusion – The Story of India's Atomic Energy Programme*, Viking, New Delhi (2002).

J. Shapiro *Radiation Protection – A Guide to Scientists and Physicians*, Harvard University Press, Cambridge, MA (1990).

ICRP *2007 Recommendations of the International Commission on Radiological Protection.* Elsevier, Amsterdam (2013).

D.V. Gopinath *Radiation Effects, Nuclear Energy and Comparative Risks*, Current Science, Vol.93, No.9 (2007).

P. D. Wilson (Ed) *The Nuclear Fuel Cycle- From Ore to Wastes*, Oxford University Press, Oxford (1996).

IAEA *Spent Fuel Reprocessing Options*, TEC-DOC-1587, IAEA, Vienna (2008).

IAEA *Radioactive Waste Management- An IAEA Source Book*, IAEA, Vienna (1992).

K.R. Rao *Radioactive Waste – The Problem and its Management*, Current Science, Vol. 81, No 12 (2001).

International Nuclear Current Issues in Nuclear Energy- Societies Council Radioactive Wastes, American Nuclear Society (2002).

Websites Nobel Lectures

World Nuclear Association

US Atomic Energy Commission

US Department of Energy

US Nuclear Regulatory Commission

US Environmental Protection Agency

UK Atomic Energy Authority

International Atomic Energy Agency

Nuclear Suppliers Group

Bhabha Atomic Research Centre Trombay

Indira Gandhi Research Centre for

Atomic Research, Kalpakkam

Nuclear Fuel Complex, Hyderabad

Uranium Corporation of India Ltd.

Nuclear Power Corporation of India Ltd.

Bharatiya Nabhikiya Vidyut Nigam Ltd

Atomic Energy Regulatory Board

Journals, Bulletins & Scientific American

Newsletters Bulletin of the Atomic Scientists

Indian Nuclear Society News

Indian Association of Nuclear and

Allied Scientists (IANCAS) Bulletin

BARC Newsletter